Lucas Grisolia

Territory, protected areas and traditional populations

Lucas Grisolia

Territory, protected areas and traditional populations

The residents of the Juatinga Ecological Reserve - RJ

ScienciaScripts

Imprint
Any brand names and product names mentioned in this book are subject to trademark, brand or patent protection and are trademarks or registered trademarks of their respective holders. The use of brand names, product names, common names, trade names, product descriptions etc. even without a particular marking in this work is in no way to be construed to mean that such names may be regarded as unrestricted in respect of trademark and brand protection legislation and could thus be used by anyone.

Cover image: www.ingimage.com

This book is a translation from the original published under ISBN 978-613-9-64641-8.

Publisher:
Sciencia Scripts
is a trademark of
Dodo Books Indian Ocean Ltd. and OmniScriptum S.R.L publishing group

120 High Road, East Finchley, London, N2 9ED, United Kingdom
Str. Armeneasca 28/1, office 1, Chisinau MD-2012, Republic of Moldova, Europe
Printed at: see last page
ISBN: 978-620-7-78164-5

ACKNOWLEDGEMENTS

I'm grateful for the opportunity to have studied Geography, a science that has contributed to my values and shaped me as a citizen.

My special thanks go to Professor Marília Steinberger, who guided me and encouraged me to pursue this subject.

I would like to thank my colleagues from my internship at the Chico Mendes Institute for Biodiversity Conservation, Tânia Maria de Souza, Mackinley Lobato de Souza and my colleague, also a geography student, Alice Vogado, for their support.

My special thanks go to Professor Doris Sayago for her participation in the defence panel, as well as her final guidance on formatting and structural adjustments; and to Carlos Felipe Abirached, also for his participation in the defence panel and for his support in the exploratory research.

I would like to thank all my long-time friends, who to this day support me in my decisions.

My special thanks go to my mum Maria de Nazaré, my dad César, my brother Vitor and my aunts Tereza and Isabel, who always supported me in whatever I needed. I would also like to thank my partner Gabriela for her patience and promptness.

SUMMARY

This monograph presents an analysis based on the geographical concept of territorial interpretation. It analyses the extent to which the creation and implementation of a conservation unit contributes to maintaining the culture of traditional populations. The case of the Juatinga Ecological Reserve, located on the south coast of Rio de Janeiro, in the municipality of Paraty/RJ, was chosen. It was necessary to study the different conceptions of territory, the history of the ideal of conservation that resulted in the creation of protected areas and the Brazilian legislation on conservation units. In addition, it was necessary to characterise the social actors present in the Juatinga Ecological Reserve: the traditional caiçara population, the environmental management body responsible and the environmental managers working in the area. An exploratory investigation of the reserve was also carried out. We can conclude that the various interests present in the creation and implementation of the conservation unit directly interfere with the maintenance of caiçara culture.

Key words:

Territory, conservation units, traditional populations, caiçaras, conflict.

SUMMARY

INTRODUCTION

The municipality of Paraty, located in the extreme south of the state of Rio de Janeiro, lies in a coastal area under the dominion of the Atlantic Rainforest. Founded in the 17th century, it occupied an important place in colonial trade. Sugar cane cultivation was the first established economic activity, and ruins of mills from that period still remain today. It then became a centre for the export of gold from Minas Gerais, which was transported along trails previously used by Indians. In the mid-19th century, Paraty exported a considerable amount of coffee, tobacco and brandy by sea. However, with the construction of the D. Pedro II Railway and the abolition of slavery, the basis of local monoculture, its decline began. The population, isolated during this period, then began to live off the incipient commercialisation of fish and subsistence farming.

In this context, the local population, generically known as "caiçara", is the result of ethnic-cultural miscegenation between the Indians, the Portuguese colonisers and, to a lesser extent, black Africans. To ensure the reproduction of their way of life, their techniques, symbologies and rituals are passed down through generations. It should be emphasised that the economic activities of the caiçaras are based on a deep knowledge of the local ecosystem, which is reflected in the development of strategies for the use and management of natural resources. This dependence on nature and natural cycles, their economic organisation with little or no capital accumulation and the fact that they don't use salaried labour classify them as "traditional populations". They practise subsistence activities for the most part, although the production of goods is present, which implies a relationship with the market. They attach great importance to their culture, symbologies and myths associated with their activities and their territory. Although this classification is controversial, as it is known that societies' relations with the territory do not remain static, the term was coined to differentiate them from modern Western society, which has a "harmonious" relationship with nature.

On the other hand, the Atlantic Forest biome, located along the Brazilian coastline, has been intensely degraded since the beginning of Brazil's colonisation. It was here that the country's first cities were established, as well as the various economic activities demanded by the international market: the extraction of brazilwood, the sugar cane cycle and the coffee cycle. Later, industrialisation, livestock farming and extensive monoculture agriculture also contributed to this degradation.

As a result, the preservationist outlook and conservation policies are strongly present in the few remaining fragments of the biome. Conservation Units (CUs), management and land-use planning instruments, are part of these policies aimed at protecting natural resources and biodiversity. The creation of protected areas seeks to regulate society's relationship with the environment.

In 1992, a state decree created the Juatinga Ecological Reserve in the south of the municipality with the aim of preserving the Atlantic Forest biome. Although the law creating the reserve mentions the importance of caiçara culture, the application of restrictive legislation regarding the use of natural resources has repercussions on the community's ways of appropriating the territory. The UC geographically delimits an area where laws will determine behaviour within it, aimed at restricting mobility and the use of natural resources. However, this form of appropriation of a territory overlaps with the appropriation of the caiçaras, who practice it with their low-impact economic activities and the reproduction of their culture. Thus, a socio-environmental conflict emerges, generated by different interests in the same territory.

Prior to the creation of a conservation unit, socio-environmental diagnoses, studies of the potential for preservation, public hearings and political articulations take place. These involve the interests of various social actors from the process of its creation until after it has been implemented, often being used to favour one group. The pressure to designate the most interesting way of using this territory is constant. In this context, traditional populations are strong allies of environmental conservation, adding their culture to the protection of Brazilian biomes. However, in order for this intimate relationship with nature to result in the conservation of their culture, it is necessary that the policies for creating protected areas take into account the interests of these communities. This monograph therefore aims to analyse the extent to which the creation and implementation of protected areas contributes to maintaining the culture of traditional populations.

In order to carry out the research, it was necessary to characterise the social actors present in the Juatinga Ecological Reserve, survey the Brazilian legislation that regulates the process of creating conservation units, together with the historical processes of constructing the idea of environmental conservation and traditional populations.

The case analysed highlights the dispute over territory. For the caiçaras, territory is closely linked to their identity. It contains social representations, myths and symbols that characterise them as a unit. On the other hand, the Juatinga Ecological Reserve delimits and controls the same territory, seeking other objectives for its use. In this way, the approaches and conceptions of the geographical concept of territory will also be discussed, which will serve as a basis for understanding the relationships in the reserve.

Firstly, a literature review was carried out on the concept of territory and the ideas of environmental conservation and traditional populations. This was followed by exploratory research. In the municipality of Paraty/RJ, statements were taken from environmental managers and the head of the Juatinga Ecological Reserve. In Saco do Mamanguá, statements were taken from caiçaras who carried out their daily activities on Cruzeiro beach. Finally, the bibliographical review, the data collected in the investigation and the legislation relevant to the issue were brought together for a critical interpretation of the relationship between the reserve and the caiçaras in the territory.

The structure of this monograph consists of five chapters. The first aims to present the concept of territory, its geographical approaches and conceptions, in order to provide a basis for analysing the relationships between social actors. The second looks at the relationship between man and nature, the idea of environmental conservation and protected areas, seeking to present the historical process of building conservation units. The third presents the idea of traditional populations, taking a descriptive approach to their characteristics and activities in order to understand how their territory is appropriated. In this way, the country's fishing activity is addressed in relation to caiçaras communities, describing their fishing and farming techniques, hunting and gathering, territorial mobility and the history of their populations. The fourth chapter presents a case study of the Juatinga Ecological Reserve, giving a general description of the region, the history of the municipality of Paraty and the context in which the reserve is located.

Finally, the fifth chapter presents the exploratory research carried out in the region, which includes statements from the environmental managers who work in the PAs, a visit to the caiçara community of Saco do Mamanguá and an analysis of the documents collected in the field.

The way of life of the caiçara communities is one of the cultural pluralities present in Brazil. Their traditional knowledge, economic activities and social organisation are different from urban-industrial society, which is why promoting the maintenance of their culture is important.

However, these communities' relations with other social actors often take a different direction. There is enormous interest from private enterprise in their territories. On the one hand, the fishing industry, which has been modernising and showing promise for the country's economy. On the other, the tourism market, which is increasingly interested in its beaches. In this way, the caiçara have been dispossessed of their land and their way of life to make way for other, more profitable economic activities.

Another relevant fact is the creation of Conservation Units as an instrument that guarantees these communities the reproduction of their way of life in their territory. However, there are also various interests behind the creation and implementation of Conservation Units and their management plans are often not adapted to the reality of the caiçaras, generating conflicts. This monograph is therefore motivated by the right to territory of traditional populations and the potential that environmental conservation has to guarantee this right.

The municipality of Paraty has overlapping land-use planning instruments. There are 5 conservation units of different categories in its territory, totalling more than 80% of its area, in addition to the municipal master plan. In this context, the aim is to analyse only the relationship between the Juatinga Ecological Reserve, managed by INEA, and the caiçaras who live within it. The hypothesis is that the relationships between all the actors involved in this context contribute to the legal gap and collusion with the reserve's private interests, directly interfering in the maintenance of caiçara culture.

CHAPTER 1

The Concept of Territory

"There is no simple concept. Every concept has components, and is defined by them"
(DELEUZE and GUATTARI apud SAQUET, 2010).

By addressing the concept of territory in this chapter, the aim is to provide a theoretical basis for understanding protected areas for conservation and their importance for traditional populations.

The concept has been worked on by researchers from various fields of study, mainly in the humanities and natural sciences, resulting in approaches and conceptions of territory that vary according to the philosophical and theoretical-methodological foundations used by the authors. For Geography, it has emerged as one of the central concepts for the renewal of geographical thinking, seeking to break with the merely descriptive approach of Classical Geography and bring a critical analysis of the humanised world.

It should be emphasised that there are two perspectives for analysing territory, according to geographer Rogério Haesbaert (2002). One is the materialist analysis, which sees territory as strongly linked to physical space, to the land. It approaches it by emphasising its physical characteristics, such as its use for human survival and reproduction. For thinkers from this perspective, human consciousness is a product of material mechanism. Another is the idealist analysis. It prioritises the symbolic character of territory, showing that myths and representations relating to the space in which individuals in a society live define it as such. As such, territory would be a subjective conception of each individual (HAESBAERT. R, 2002).

The concept of territory is also closely linked to the concept of space. Both are discussed together by authors who contribute to the renewal of geographical thinking. However, it is not necessary to discuss the concept of space here, but in order to approach territory, it is necessary to mention it. Space, then, according to the definition given by geographer Milton Santos, "is *the indissoluble union of systems of objects and* systems of actions, and their hybrid forms, the techniques, which tell us how the *territory is used (...)"* (STEINBERGER, 2006: 60).

To begin with, R. Haesbaert, in his book "O mito da desterritorialização: do fim dos territórios à multiterritorialidade" (The myth of de-territorialisation: from the end of territories to multi-territoriality), presents a summary of the notions of territory in a didactic and interesting way, which we can take as a starting point for other approaches and conceptions of the concept. He groups the notions into three basic strands.

The first, the political, which refers to power relations in general, or juridical-political (also referring to all institutionalised power relations): this is the most widespread, where territory is seen as a delimited and controlled space, through which a certain power is exercised, most of the time, not exclusively, related to the political power of the state.

The cultural or symbolic-cultural approach prioritises the symbolic and more subjective dimension in which territory is seen as the product of a group's appropriation/valuation of their lived space.

The economic emphasises the spatial dimension of economic relations, territory as a source of resources and/or incorporated into disputes between social classes and in the capital-labour relationship, as a product of the territorial division of labour, for example.

These three strands are closely linked to the human sciences. However, we can add another strand before them, more closely linked to the natural sciences, the naturalist strand. This uses the notion of territory as the basis of society-nature relations and brings the idea of a "natural" behaviour of man in relation to his physical environment. Human territoriality would be the instinct to look after the natural resources on which they depend for survival.

This biological notion of territory led neo-Darwinian author Robert Ardrey to write:

> *We act the way we do for reasons of our evolutionary past, not our cultural present, and our behaviour is as much a mark of our species as is the shape of our thigh bone or the configuration of the nerves in an area of the human brain. (...) if we defend the title to our land or the sovereignty of our country, we do so for reasons no less innate, no less inextirpable than those that make the owner's fence act for a reason indistinguishable from that of its owner when the fence was built.* Man's territorial nature is genetic and inextirpable (HAESBAERT, 2007: 46).

Naturalist authors saw territory as having defined boundaries and as a source of basic resources for life, defended by a group as an exclusive reserve. They saw man as a natural being, with impulses to possess and defend his territory. In this way, such behaviour would be the same in any society.

These theses have been strongly criticised in the academic world, including by biologist W. H. Thorpe (1974), who highlights the risk of this analogy by giving space to justify human aggression to the predatory nature of our species. However, other scientists, such as anthropologist José Luis Garcia, point out that the total separation of man and nature can lead to extreme anthropocentrism. He therefore warns against the inappropriate application of Darwinist conclusions to the human world, but believes that total separation is, at the very least, reckless (HAESBAERT, 2007).

The direct analogy of characteristics from the animal world to the human world superimposes genetics on society as the driving force behind individuals' attitudes, and is therefore highly questionable. Of course, there's no denying that territorialisation means taking care of the perpetuation of a species or social group. However, what is contested is the principle that leads to this. When, for example, the Landless Movement in Brazil today claims land for the use of its class, it is seeking to territorialise itself. The social and political context in this case leads them to fight to continue to reproduce their economic activity and peasant way of life, rather than a genetically determined impulse.

Friedrich Ratzel, a biologist who entered the field of geopolitics in the context of the unification of Germany in the 19th century, understood territory as an area and natural resources (soil, water, climate) and the state as the main form of power and control over it. "The *society we consider, whether large or small, will want to maintain possession of its territory above all else, on which and thanks to which it lives. When this society organises itself for this purpose, it becomes a state"* (RATZEL, 1990 apud SAQUET, 2010, page 30). Territory, then, means the appropriation and domination of natural physical bases by a politically organised people, in other words, by the state.

Even though Ratzel tried to work out the relationship of power in the territory, he still limited it to a simple receptacle for human activities, a substrate/stage for the realisation of life, synonymous with soil/land and other natural conditions, fundamental to all peoples, both savage and civilised, under the domination of the state (SAQUET, 2010).

The inertia attributed to territory in this naturalistic notion is not present in the other notions. The cultural, political and economic components of the concept reveal it as an active element in society and influential in the lives of individuals. Therefore, in its political notion, it is essential to understand its intimate relationship with power.

Saquet, in his book "Approaches and conceptions of territory", presents power as inherent in social relations. Power would be present in the actions of the state, institutions, companies, in short, in social relations that take place in everyday life, aimed at control and domination over men and things. Thus, when quoting Michel Foucault, he explains that power can be understood as "the Power", a set of institutions and apparatuses that guarantee the submission of citizens in a given state, and power, with a small letter, which would be the multiplicity of relations of force inherent in the field in which they are exercised. As such, power is produced in the field of relationships, at every moment. *"It is not an institution, but the name given to a complex situation of life in society*

(SAQUET, 2010: 32).

Anthropologist Pierre Clastres, in his book "Society against the State", argues that there is no such thing as society without power, since it is immanent to society. He also points to two main modes of power: coercive power and non-coercive power. The former manifests itself in command-obey relationships and violence. It is characteristic of power centralised in social groups or classes, where there are unilaterally imposed rules. This power, coming from the hierarchisation/differentiation of a social group or class, would be the embryo for the formation of the state. The second manifests itself when power is dissolved in society, without being concentrated. It emanates from relations between individuals.

Therefore, power is not something that is acquired, but exercised. It is internal to economic, cultural and political processes. Power means heterogeneous, variable social relations with diverse intentions; relations of force that go beyond the actions of the state and involve and are involved in other processes of everyday life, such as the family, university, church and the workplace. Territory, in this multidimensionality of the world, takes on different meanings, based on plural, complex and unified territorialities. Thus, the meanings of territory change as the understanding of power relations changes (SAQUET, 2010).

To understand the political notion of the concept, it is important to emphasise that power relations take place between social actors and the territory. In Ratzel's geopolitics, the state was the only actor, assuming that there were no internal subdivisions or contradictions. However, for the purposes of analysing the concept, the state is seen as another actor within a plurality of actors who relate to the territory in pursuit of their interests. The state itself can also be seen as the representative of various actors.

For Robert David Sack (apud M. A. SAQUET, E. B. SOUZA, 2009), for there to be territory there must be a delimitation of area, control and a form of power. Its effects and results depend on who is controlling whom and for what purposes. The border takes centre stage in his approach insofar as it makes control possible. In this way, territoriality would be a strategy of the individual or social group to establish limits of access to areas, objects and people. It must involve classifying areas, in the sense of who it belongs to; it must contain a form of communication, such as a mark or sign; and it must also involve access control actions.

Territory, for Claude Raffestin, would be an asset for power to be exercised, "it is a *particular asset, resource and obstacle, continent and content, all at the same time. Territory is the political space par excellence, the field of action of power"* (RAFFESTIN, 1993 apud M. A. SAQUET, E. B. SOUZA, 2009 page 40). The French author takes a relational approach to the concept, where power is fundamental to its understanding. The relationships between actors are not bilateral, but multilateral, as they are receivers and transmitters of energy and information flows. For Raffestin (apud M. A. SAQUET, E. B. SOUZA, 2009), energy is the potential that enables the displacement or modification of matter, and information is form or order in all matter or energy. In this way, any social actor is part of a system of multi-scalar interrelationships (global, regional and local) made up of political, cultural and economic elements that are dynamic on a timeline. By acting on behalf of their interests, they are intermediating a flow of energy and information and, at the same time, producing it.

He also emphasises the inseparability of the concepts of space and territory. For him, *"space is as if it were a raw material, prior to territory, that is, from it and from actions of appropriation, led by actors, the territorialisation of space occurs. Space becomes a product"* (M. A. SAQUET, E. B. SOUZA, 2009). Territoriality is multidimensional and inherent to life in society. Man experiences social relations, the construction of territory, interactions and power relations in his daily life, thus revealing the construction of meshes, knots and networks. It has material and immaterial components, such as technologies and religion. It would be expressed in the symbolic mediation that the materiality of places exerts on social action (SAQUET, 2010).

Territory, then, is a fragmentation of space resulting from its control, domination and appropriation. It has materiality in the physical forms that make it up, both artificial and natural. It has immateriality in its flows of information and energy, and in the symbolic value it has for society. Territoriality is a behaviour linked to the techniques and actions of a given society, which thus differentiates the space in which it lives. It is effected by social relations, territorial identity, a sense of exclusivity and the compartmentalisation of human interaction in space (SAQUET, 2010).

There is also the territory conceived from the concept of frontier and pioneer fringe. The border is analysed regionally as an area of transition and transnational relations. There is delimitation and demarcation; economic and political factors; the jurisdiction of the state as the central power. The fringe corresponds to an area or zone of social occupation; it is a phenomenon that does not depend on the power of the central authority of the state, but on forces that act multilaterally. Territory is thus the result of the definition of borders and frontiers of occupation and settlement, inherent in the actions of the state (legalisation, inspection and control) and other actors (SAQUET, 2010).

Territory is understood to be an area where there is an element of centrality. This can be an actor exercising sovereignty over people or over the use of a place. Jurisdiction and policies are attributes that permeate society as a whole and give the state sovereignty, making it a central element at this point. On the other hand, the predominance of one form of appropriation of an area (by a company or a traditional population) also shows a centrality in the territory, shaping its artificial physical aspects. Therefore, if you look at a large industrial estate or an artisanal fishing community, you can see a centrality in the use of the territory.

Rogério Haesbaert, in his book "Território alternativo" (Alternative Territory) (2002), when talking about the economic notion, explains its links with the naturalist notion by addressing territory in relation to its material aspects and its functionality. It would be a vital resource base for human survival and material reproduction. However, it does not consider it inert. Its material and immaterial characteristics define its use and economic function, which are central elements of this approach to the concept. Territory is defined by the way it is used for production.

Especially from the 18th century onwards, when the Industrial Revolution and the Bourgeois Revolution took place in Europe, the ideology of the free market and private property promoted a territorial rearrangement. The old regimes were overthrown in favour of a state of legal equality, laying the foundations for industrial economic growth. Peasants are driven off their land to become proletarians. Agrarian reforms were promoted to change the use of the land to meet bourgeois demands. The countryside and the city take on new meanings. In this way, there is a constant reorganisation of the territory by the agents of capital, who in turn are inseparable from the modern state. For the American geographer Robert David Sack (apud HAESBAERT. R, 2007), private property, because it is legally recognised, is one of the most familiar forms of human territoriality.

For Milton Santos (apud HAESBAERT. R, 2007), use defines territory par excellence. However, territory is not used uniformly. There are sections of territory that acquire different functionalities throughout the historical process of their formation. When you look at the objects, actions and techniques present, you can see their function. He works with the concept of "used territory", which is the result of the interaction of systems of objects and actions, and is therefore synonymous with geographical space. Therefore, "the actions of *actors using the territory over time produce space as a totality that is expressed materially in a territorial configuration*[1] (STEINBERGER, 2006: 62).

Strongly influenced by historical materialism and Marx's theory of value, this economic notion is presented as the social organisation of territory by the productive forces and relations of production of capitalist expansion. Thus, authors such as Francesco Indovina and Donatella Calabi (1974) present the capitalist use of territory in terms of the process of production and the extraction of surplus value; the circulation and valorisation of capital; and the reproduction of the labour force. For both, the first use of the territory is to facilitate the extraction of surplus value through its productive location and infrastructure (for financial, administrative and commercial purposes). The next use occurs in the circulation and reproduction of capital on the market. The consumption of goods and/or services is profitable when the territorial bases are favourable, as is the case with cities, where individual workers and consumers are concentrated (SAQUET, 2010). In this way, the territory is approached as a use value for the reproduction of capital. Both its physical character, such as the natural environment and infrastructures, and the flows of information, goods and people, are attributes that influence the territorialisation of capital.

This notion of territory is closely linked to multiscalarity in terms of the relationships between social actors. With their political, economic and cultural characteristics, social actors relate to each other, forming connected and articulated territories. The idea of continuous territories connected by networks (such as roads, maritime routes and communication networks) brings with it the notion of fluidity. Both in global relations, such as between nation states or economic blocs, and in local relations, such as between a town hall and a residents' association, the articulation between territories is taking place.

These relationships also take place across scales. An environmental policy, for example, which is discussed at international conferences between states, is later applied in the context of local populations. Thus, for Claude Raffestin, there is a construction of territorial meshes, nodes and networks: *"Networks, central to the production of territory, are central to his proposal for a non-areal territorial approach; they are understood through the existing complementarity between circulation and communication, as material and immaterial flows"* (SAQUET, 2010: 76).

The contemporary world is experiencing a multiplicity of scales. As a result, there are multiple territories, thus strengthening the use of the term "territory-network", which contributes to understanding these articulations between forms of territorial appropriation. This term arises from the techniques created for the use of territory, such as transport networks and communication networks. This gives rise to the notion that territories are permeated by flows of goods, people, information and capital (HAESBAERT. R, 2002).

Of course, this network is not homogenous. There are nodal points where flows converge, as well as territories with a higher density of flows than others. Therefore, the appropriation/domination of territory also varies according to its geographical position in relation to the network. This, in turn, is very dynamic and can be remodelled, advancing or retreating in the territory.

> It is true that, in a certain sense and under certain conditions, there are effectively global networks, involving the world as a whole. But since one of the characteristics of networks is that they only form lines (flows) that connect points (poles), never filling the whole of space, there are many interstices that offer other ways of organising space. Identifying networks with a planetary dimension which, according to some authors, serve as an embryo for the formation of a world-

This approach of territories articulated in networks brings the idea that there are multiple interests in the form of their appropriation/domination. Therefore, a characteristic of the concept that deals with the dynamics of territories is the so-called deterritorialisation.

> Concrete reality involves a permanent intersection of networks and territories: more extroverted networks which, through their flows, ignore or destroy borders and territories (and are therefore deterritorialising), and others which, due to their more introverted nature, end up structuring new territories, strengthening processes within the limits of their borders (and are therefore territorialising) (HAESBAERT, 2002: 123).

For philosophers Gilles Deleuze and Felix Guattari, "the *hominid: from the moment it is born, it deterritorialises its front paw, it pulls it out of the earth to make a hand out of it, and reterritorialises it on branches and tools. A stick, in turn, is a deterritorialised branch (...)"* (M. A. SAQUET, E. B. SOUZA, 2009: 26). There is thus a constant movement of change, in which people and objects are deterritorialised and, as part of the same process, reterritorialised.

At this point, based on the networked world, a strong deterritorialising agent is capital. The logic of market expansion that flows through these networks reaches territories that have been appropriated/dominated in other ways, with the aim of territorialising them. In opposition, less politically organised social groups are forced to deterritorialise. These are processes that demonstrate the fluidity and overlapping of territories in power relations that transcend the global, regional and local scales.

Within the discussion of the concept, there are two terms widely used to designate the formation of territories through social relations: domination and appropriation. To clarify the meanings, Haesbaert (2002) states that domination tends to give rise to purely utilitarian and functional territories, linked to a use value. Appropriation, on the other hand, brings the idea of symbolic value, a relationship of identity with the territory, of a cognitive nature. However, both are always present: *"to associate physical control or "objective" domination of space with symbolic appropriation, which is more subjective, implies discussing territory as a space that is simultaneously appropriated and dominated, that is, over which not only physical control is built, but also bonds of social identity"* (HAESBAERT, 2002: 121).

Therefore, when we talk about domination, we infer a materialist approach to territory. Access control, the delimitation of areas, the segmentation of space and the definition of its use appear as the main axis of political and economic notions. When it comes to appropriation, there is an idealistic approach to the concept. Territory enters the field of representations, symbols and myths. Every society is cultural, and so is the way it relates to its surroundings.

The cultural notion emphasises that everything that surrounds man is endowed with meaning. The feeling of belonging to the territory implies the representation of cultural identity. Territories are not thought of as polygons, but as a set of meanings. As such, this notion is best worked out on a local scale. Places, routes and objects have an intimate relationship with individuals. They are part of people's daily lives. They are used on the basis of culture (HAESBAERT, 2007).

For Bonnemaison and Cambrèzy (apud HAESBAERT, 2007: 72):

> The power of the territorial bond reveals that space is invested with values that are not only material, but also ethical, spiritual, symbolic and affective. This is how cultural territory precedes political territory and even more so precedes economic space (...) territory is not just about function or having, but about being. To forget this spiritual and non-material principle is to fail to understand the tragic violence of many struggles and conflicts affecting the world today: to lose your territory is to disappear.

The territory approached through the cultural notion can be analysed in any society. In modern urban-industrial society there is a cultural relationship between the consumption of space and the consumption of ways of life. In large cities, people live in territories connected globally by information and consumer culture. In this sense, the way of life in urban societies is a result of the fluidity present in their territory. In so-called traditional societies, such as the caiçaras, the way of life is permeated by the influences of local networks. Of course, the global network expands, promoting its influence on these populations. These populations, organised in local networks, also promote their influence on global networks.

As a conclusion to this presentation of the notions of territory, it is worth adding that neither is absent from the other. Thinking of a territory from an integrative perspective would be the way to encompass its characteristics as a whole. It is inextricably linked to a symbolic or cultural dimension and a material, economic and political dimension. Territory is administration, organisation, control and use. It is where power relations take place between actors with different interests. It is also identity, affectivity, a space for reproducing a way of life, symbologies and social representations. As Haesbaert summarises, "it is the *product of an unequal relationship of forces, involving political-economic domination or control of space and its symbolic appropriation, sometimes combined and mutually reinforced, sometimes disconnected and contradictorily articulated"* (HAESBAERT, 2002: 121).

Territory and Conservation Units

The growing interest in environmental conservation, resulting from a revaluation of the natural environment, has created instruments that aim to standardise or control a certain way of using the territory. The creation of conservation units delimits a space that will be administered by a state body that, through policies and laws, aims to restrict access and determine activities in favour of maintaining biodiversity.

Of course, when it comes to delimitation and access control, the territory of these protected natural areas has a materialistic character, since they are secluded areas with specific uses and functionalities. With nature gaining market value and being associated with quality of life, there is growing interest in developments that provide this new function for consumers. On the other hand, these are not empty territories. They are appropriated by local populations. Thus, not looking at their idealistic character, in which these populations have a relationship of identity with the area, has generated conflicts of overlapping territories.

Therefore, as a result of international discussions and the social desire for an environmental balance, various models of protected areas have been created, with different objectives: protecting the habitats of endangered species; preserving areas of great beauty and high biodiversity; conserving areas where man's degrading occupation poses risks to society; preserving natural environments for tourist use; and prioritising the activities of populations that use biodiversity for their livelihoods.

The relationship between social actors such as the state, organised civil society, NGOs and social movements, in conjunction with conservationist interests, has produced territories that are dominated/appropriated with the centrality of maintaining biodiversity and traditional practices. Today, there are territories where cultural appropriation is valued and where the parsimonious use of natural resources is backed by federal legislation.

There is also the view that these areas cannot be enclaves of conserved ecosystems surrounded by degraded environments. There must be articulation between them with the aim of physically connecting these territories so that genetic flow can take place, providing the greatest possible variability of genes. Articulation in information networks is also important in terms of constantly improving the implementation and management processes of these areas and connecting them to nodal points of knowledge production, such as universities and research centres.

To conclude, it is important to understand that conservation units, as they stand today, derive from a process of historical construction of ideologies and instruments for managing the natural environment. Today, they are involved in political articulations that take into account the numerous interests of the social actors present in the territory, at various scales. They have an economic character in terms of the revalorisation of nature, which becomes a promising market that attracts capital to the technology production and tourism sectors, for example. Culturally, they are appropriated by different social groups who in their daily lives have intimate relations with the territory, making it an inseparable part of their identity.

The aim here, therefore, is not to understand the singularities of conservation units as unique and exceptional cases, approaching them in a descriptive manner, but to investigate the processes and social relations that permeate their creation and management. In this way, we can provide a broad notion of environmental conservation, helping to see protected areas within historically constructed processes.

CHAPTER 2

Environmental Conservation and Protected Natural Areas

> *"Advances in new technologies, by making it possible to understand the economic importance of biodiversity, have aroused the interest of large economic segments in controlling its exploitation, as a way of generating knowledge and new possibilities for its commercial use."* (GUERRA and COELHO, 2009: 27).

Nature is a social concept. By politically and culturally determining what nature is, human beings are transmitting to the physical and biological environment the characteristics of the society in which they live. Nature does not exist separately from man in this day and age, where technological advances have led to extensive knowledge of the earth's surface. There is no part of the planet that humans don't know about. Even the environment that has never been accessed *in situ* is permeated with laws that condition its appropriation. For example, the hard-to-reach interior of the Amazon rainforest, although uninhabited, is under a framework of laws and political articulations that determine its preservation or exploitation. Nature is conditioned by the social aspects in which it is inserted.

In this context, nature is part of a capitalist society where the search for new markets is necessary for this economic system to reproduce itself. The opening of eyes to the reality that natural resources are finite and that large-scale industrial production would have to be rethought has given rise to a promising market, the "green" market. This market follows the logic of accumulation, which sees nature as a scarce resource with the potential to generate capital, fundamentally in terms of the use of biodiversity conditional on the advancement of technology (BEKER. B, 2005).

The realisation of the economic importance of biodiversity highlights the need for an international commitment to its conservation, bringing in rules and regulations for its use. As a result, developing countries with a high level of biodiversity, such as India and Brazil, have entered into international discussions to define how their resources will be used, highlighting an even bigger discussion: the role of sovereignty for states with megadiversity.

Major technology producers such as Germany and the United States find a vacuum for their products in the biodiversity of the third world. The unequal distribution of technology and natural resources generates a flow of economic interests towards peripheral countries, where ecosystems are being degraded by extensive farming and mining. Thus, a reality becomes clear in which the country's economic activities are geared towards meeting the demands of the international market, opening up another market to be exploited by those who own the technology.

The interests surrounding the natural environment stem from its revalorisation. Its preservation is linked to a new market, but also to the social and economic consequences that its degradation can bring. Therefore, in order to understand society's relationship with the natural environment and the creation of protected areas for conservation, it is necessary to understand the historical processes of constructing ideologies and forms of appropriation/domination of the territory that result in the environmental policies applied today.

2.1 Conservation

Man's relationship with nature conservation is cultural, political and economic. At the beginning of the development of human societies, when man has to guarantee the survival of his species, he needs to feed and protect himself. To do this, he makes use of the natural environment, choosing the main elements that provide him with his basic needs and conserving them. However, what is basic varies for societies according to the historical moment they are living in and their social relations. For example, for a society that is at war, basic resources are chosen to be exploited for this activity. In this case, conservation has a material, use-value character, but in addition to this, it also has a symbolic value character.

There are reports (FERRY apud VIANNA, 2008) from the time when human actions were inspired by the Bible, in which cases of animals damaging farmers' crops were taken to court and they won. The argument was that since they are also God's creatures, they have the right to subsistence, and it was not up to man to deprive them of it.

During the Renaissance, at the end of the 13th century, reason and science were valued. Anthropocentrism, according to Viana (2008), was the origin of the separation in the relationship between man and nature. The notion of civility came to exist in opposition to animality or any connection with the natural world. This feeling of superiority led to the idea that civilisation was mastery over the natural order. Then, with the evolution of the human and natural sciences, anthropology identified man as the only being endowed with culture, distinguishing him from other species.

In the natural sciences, the aim was to study natural phenomena, understood as those that manifest themselves with their own dynamics, without human participation. Nature was thus defined as something external to man. Later, this separation was reinforced by urban-industrial society, which saw nature as a source of resources for economic development.

Since the Industrial Revolution, technology has developed very quickly, as has the growth of cities. Industrial society began to appropriate the environment in favour of the reproduction of its economic system, jeopardising its very maintenance. As a result, resources such as soil, water and forests are exploited to the point of exhaustion. In this way, they gain market value, from stock to use, while the consequences of their overexploitation are shared by society.

"The *profits from this exploitation are individualised, but the damage caused by the irrationality of its use - extinction of fauna and flora, pollution of soil, water, air, etc. - are socialised*" (VIANNA, 2008: 138).

During this period, the idea of environmental conservation linked to quality of life began to emerge. People living in the city began to notice the unhealthiness of the environment due to polluted air and poor sanitary conditions. The countryside began to be frequented to escape urban life, and thus nature began to be valued as synonymous with beauty and purity.

The way in which society was organised brought with it countless problems. It needed to be challenged. Ideologies emerged to rethink the way society relates to the environment, ranging from moderate reformism to ecological radicalism. Thus, the realisation that nature is finite was the basis for today's concept of conservation.

Vianna (2008), when talking about the idea of conservation in his book "From invisibles to protagonists: traditional populations and conservation units", quotes the author Luc Ferry. For Ferry, the currents of conservationism that have emerged from protest fall into three broad categories: anti-modern, modern reformist and modern (FERRY apud VIANNA, 2008).

The first condemns anthropocentrism and modern humanism, proposing that nature has value in itself. They believe that man is a natural being and that other species are of equal importance in the world. Part of this current is deep ecology, which emerged in the 1970s, preaching that the intrinsic value of nature should be protected by legal rights.

The second, a reformist view of modernity, is aware that exacerbated anthropocentrism has caused damage to the environment, although there are ways out within the context itself. With the veneration of nature as spiritual nourishment and as an aesthetic and recreational attraction, man began to consume it without questioning his subordination, still stimulated by the ideology of progress. The idea of rationalising natural resources emerged, with the understanding that if the destruction of the environment continued, man would put his very existence at risk.

The latter continues to be driven by technical-instrumental reason and is insensitive to environmental issues. Extreme anthropocentrism, which continues to see nature as a storehouse of resources to be exploited, would circumvent any future problems that humanity might face with technological advances.Therefore, by revaluing nature, modern society has developed legal and institutional instruments to protect it. The degradation of natural environments has caused social and economic consequences that can be avoided through the rational management of natural resources, with a view to maintaining them for future generations. This makes it necessary to structure institutional bases that can deal directly with the environmental issue.

2.2 Protected Natural Areas.

> *"A protected area is a clearly defined geographical space, recognised, dedicated and managed, by legal or other effective means, to achieve the long-term conservation of nature, associated with environmental services and cultural values."* (International Union for Conservation of Nature, 2011).

Through various forms of relationship between man and nature, the environment has been greatly altered. The challenge of survival has shown man throughout history that certain changes to the environment in which he lives can jeopardise the existence of his species. For this reason, areas that manage the use of resources in a parsimonious way, with a view to maintaining them for the future, have existed in society since its earliest days, with differences related to the cultural, economic and political context in which it is inserted.

There are reports highlighting the desire for environmental conservation in societies that existed in Asia and the Middle East:

> *In India, 400 years before Christ, all forms of use and extractive activities were prohibited in the sacred forests; 700 years before Christ, Assyrian nobles established hunting reserves, similar to the hunting reserves of the Persian Empire in Asia Minor, established between 550 and 350 years before Christ; in China, protective laws were established for wet plains during the sixth century after Christ; Venice created deer and wild boar reserves at the beginning of the eighth century; in Brittany, forestry laws were enacted in the eleventh century* (DAVENPORT and RAO apud GUERRA and COELHO, 2009: 31).

Conservation, therefore, was focussed on specific resources that were of interest to the segment of society with decision-making power in the territory. There was therefore no legislation governing their creation, they were unilateral impositions. From the end of the 19th century and the beginning of the 20th century, the aim of creating these areas became to preserve the landscape and witness wild nature for future generations. A confrontation arose between the idea of preservation and conservation. The latter practises the moderate use of natural resources in order to maintain them. The former defends unspoilt wilderness by preventing direct contact with man.

According to Vianna, the first legally established protected area was created in the United States in 1872 - Yellowstone National Park. As a result of the preservationist vision of the United

States, which had seen its natural beauty devastated by rapid economic growth, the park was intended to keep that remnant of wildlife untouched for public contemplation and for future generations. Nature was therefore seen as the antithesis of development and had a strong commercial appeal, being linked to a sense of freedom and purity. The same model was subsequently used in several countries such as Canada, Australia and New Zealand.

A policy of protected areas was begun with the aim of separating natural areas considered unspoilt from the development of extensive agriculture and industry. However, the parks created only allowed indirect use of resources, recreational visits and scientific research. The presence of traditional peoples in these areas was disregarded, as they were removed from their territories to give way to the idea that protected nature is nature without man. The Wilderness Act of 1964, which defined wilderness areas as those that had not been subjected to human action, reflects the idea of untouchability where man is a visitor and not a resident (VIANNA, 2008). In this way, parks were conceived as recreational and scientific resources for urban man.

On the other hand, in Europe, protected areas began to be created with the influence of scientists, with the aim of protecting certain endangered species. As the concepts of ecology and ecosystems were consolidated, the objectives of conservation were broadened to include more than just specific species, but also habitats. However, the countries of the continent adopted a conservationist approach to environmental protection, given that a large part of their territory was already occupied by urban and rural populations (Ibid).

The importation of the preservationist model of protected areas by peripheral countries has been criticised. As Larrère and Larrère (MEDEIROS apud GUERRA and COELHO, 2009, page 33) *cite, "With ethnocentrism and imperialism, wildlife preservation policy is a luxury of rich and developed countries that is inaccessible in poorer countries, harming them when it is applied to them".* These criticisms have prompted reflection in the countries on the protection practices to be adopted and on the development of a model more suited to the cultural and economic realities of these societies.

The environmental issue then came to be discussed as essential for the survival of man and other forms of life on the planet. From the 1970s onwards, a phase of conflicts, discussions and multilateral agreements began at international level. The United Nations Organisation (UNO) became the main body promoting these discussions, organising international conferences.

The International Union for Conservation of Nature (IUCN) took the initiative to categorise protected areas on the basis of scientific criteria and became the global benchmark for creating national systems of protected areas. UNESCO's Man and the Biosphere Programme (MaB

Programme) creates biosphere reserves. These are areas geared towards co-operative research, the conservation of natural and cultural heritage and the promotion of sustainable development. They have appropriate zoning, defined policies and action plans and a participatory management system involving various segments of government and society. In Brazil, the Atlantic Forest, the Cerrado, the Pantanal, the Central Amazon and the São Paulo Green Belt are biosphere reserves (GUERRA and COELHO, 2009).

Today, the World Bank and the Inter-American Development Bank (IDB) are key organisations in financing conservation efforts. As a result, these organisations gain political clout in the discussions and tend to represent the interests of large corporations and developed countries. The role of these agencies is often contradictory, in that they support highly degrading projects: "The *World Bank and other international agencies finance external adjustment or stabilisation projects that involve deforestation, industrial pollution and mineral depletion* (GUERRA and COELHO, 2009: 36).

In Brazil, discussions on the environmental issue began in the 19th century. However, the political, legal and institutional apparatus only appeared at the beginning of the 20th century.

At first, the state forestry services were the executing bodies of environmental policy in the country, playing an important management role. However, it was in the 1930s that the state became sensitive to this issue. Brazil was going through a nationalist government, which aimed to transform the country from an agrarian to an urban-industrial one. For this reason, the state needed to control its natural resources more effectively. Thus, in 1934, the Brazilian Nature Protection Conference was held, one of the objectives of which was to put pressure on the government to create a national system of conservation units (SAMPAIO apud GUERRA and COELHO, 2009).

In 1937, the first national park actually established in the country was created, the Itatiaia National Park. This was followed by the Iguaçu and Serra dos Órgãos National Parks in 1939. The creation of these parks followed the preservationist model adopted in the United States. However, the North American model sought to preserve areas far from urban society, while in Brazil, areas close to dense human settlements were chosen, where their activities were threatening the integrity of ecosystems (Ibid). Therefore, these first conservation units had the objective of territorial planning to avoid possible immediate environmental degradation, being implemented in areas where there were already territorial clashes over the use of resources, thus making their management quite complicated.

Planning and the participation of local communities in the creation of conservation units were left in the background. Immediacy and unilateral decisions ended up introducing conservation units into contexts that were not prepared for this type of state intervention, generating a repulsion on the part of society towards environmental preservation.

Until before the military regime, conservation units continued to be created, most of them related to the interests of certain social groups, but without any structural evolution in the creation process. During the period of military governments, between the 1960s and 1980s, environmental policy continued to expand. However, the centralising and authoritarian nature typical of actions during this period remained,

represented the impossibility of making adequate and lasting progress on a national environmental system, even though there has been significant progress from a legal and institutional point of view (GUERRA and COELHO, 2009).

The military also contributed to expanding the categories of existing conservation units. Through decrees, biological reserves, ecological stations, ecological reserves and environmental protection areas were created, expanding the way these areas were managed and adapting to specific conservation demands. Therefore, by 1985, just under half of the conservation units in existence today had already been decreed (Ibid).

With the tendency to bureaucratise the state in order to centralise power, the military found an institutional vacuum in the conduct of environmental policies at the time. These, which had historically been the responsibility of the Ministry of Agriculture, were sparse and under the control of the segment of society interested in unrestricted land exploitation. In 1967, the government created the Brazilian Forestry Development Institute (IBDF), an autarchy linked to the Ministry of Agriculture responsible for part of environmental policy, mainly the management of all existing federal conservation units. In 1973, the Special Secretariat for the Environment (SEMA) was created, responsible for drawing up and implementing environmental policy and the body that would form the basis for the creation of the Ministry of the Environment (Ibid).

With the foundations for implementing an environmental policy in the country structured, the National Environmental Policy (PNMA) was launched in 1981. This instrument, which systematises environmental management and its guidelines, takes an important step towards structuring environmental institutions and constitutes the National Environmental System (Sisnama).

From the 1980s and into the following decade, environmental policy in the country gained momentum. With democratisation, discussion of the environmental issue was boosted by the emergence of organised groups working directly on it. As a result, in 1985, the Ministry of Urban Development and the Environment was created from the SEMA structure. Later, in 1999, it was consolidated as the Ministry of the Environment. The Brazilian Institute for the Environment and

Renewable Natural Resources - IBAMA was created in 1989 from the structures of the IBDF, which made it possible to centralise the implementation of the PNMA (Ibid). In 2007, the Chico Mendes Institute for Biodiversity Conservation (ICMBio) was created from an IBAMA directorate to manage federal conservation units.

The 1988 constitution also contains an article (Art. 225, CF/88) in Chapter VI that further strengthens the structures of Brazilian environmental policy. In order to enforce everyone's right to a balanced environment, it lays down guidelines to be followed. Two of them emphasise the importance of protected natural areas.

> *"Art. 225. Everyone has the right to an ecologically balanced environment, which is a good for the common use of the people and essential to a healthy quality of life, imposing on the public authorities and the community the duty to defend and preserve it for present and future generations. (...)*
>
> *§ Paragraph 1 - In order to ensure the effectiveness of this right, the public authorities shall be responsible for:*
>
> *I - preserve and restore essential ecological processes and provide for the ecological management of species and ecosystems. (...)*
>
> *III - to define, in all units of the Federation, territorial spaces and their components to be specially protected, with alteration and suppression permitted only by law, with any use that compromises the integrity of the attributes that justify their protection being prohibited."*

As a result, after almost ten years in Congress, the National System of Conservation Units (SNUC), Law 9.985 of 2000, was created. This system organised the management of conservation units in the country, regulating their various categories and objectives. The individualised management of each unit made it difficult to define a state conservation policy, so standardising the management of these areas also served to support the affirmation of such policies on the national stage (GUERRA and COELHO, 2009).

The creation of the SNUC made it possible for units of different categories to be managed in an integrated way and by different spheres of government. The establishment of mandatory management councils for the units, with the participation of the actors present in the area, brings social control to the heart of management. In this way, the unilateral creation of units is abandoned and participatory management by society is adopted.

The evolution of the environmental issue on the national political scene is certainly not over yet. In addition to the lack of cohesion between the Sisnama bodies and a shortage of civil servants, the lack of enforcement is a major obstacle. To this end, the National Environment Fund (FNMA) and the Brazilian Biodiversity Fund (Funbio) were created. Both are fuelled by resources generated by environmental laws, with one of their main investments being land compensation for land expropriated by the government for conservation purposes.Until the end of the 1980s, protected areas were the result of a vision of appropriation of natural resources and territorial control. The situation changed when the state incorporated a strategic conception of the environment, in which biodiversity became the central concept of conservation policy. In this way, Brazil occupies a space on the international stage where it is responsible for conserving the planet's remaining biodiversity. The creation of protected areas then incorporates the participation of civil society and also adds the cultural diversity of the Brazilian people.

2.3 Characteristics of the National Environmental Policy and the National System of Conservation Units

The main objectives of the National Environmental Policy are to make economic and social development compatible with the preservation of environmental quality and ecological balance; the definition of priority areas for government action relating to ecological quality and balance, taking into account all political and administrative spheres; the establishment of criteria and standards for environmental quality and rules relating to the use and management of environmental resources; the preservation and restoration of environmental resources with a view to their rational use and permanent availability (Law No. 6.938/81).

To achieve these objectives, the PNMA uses instruments. Thus, it adopts the creation of conservation units as an important territorial management technique for environmental protection.

> *"Art. 9 - The instruments of the National Environmental Policy are:*
>
> *VI - the creation of territorial spaces specially protected by federal, state and municipal authorities, such as environmental protection areas, areas of relevant ecological interest and extractive reserves. "* (Law 6.938/81).

Sisnama aims to transform the environmental management process into a system made up of the three political-administrative spheres of government and civil society, enabling integrated management of environmental actions and greater efficiency in conservation.

Sisnama is structured around a higher body, the Government Council, directly linked to the Presidency; an advisory and deliberative body, CONAMA; a central body, the Ministry of the Environment, with the aim of coordinating, supervising and controlling, as a federal body, the policy and guidelines set for the environment; executing bodies, such as the Brazilian Institute for the Environment and Renewable Natural Resources (IBAMA) and the Chico Mendes Institute for Biodiversity Conservation (ICMBio), with the aim of implementing environmental policies; sectional bodies and local bodies, responsible for implementing programmes, projects and controlling and inspecting activities that potentially degrade the environment, at state and municipal level.

Another important instrument of Brazilian environmental policy, created during the military regime, was the National Environmental Council (CONAMA). This is an advisory and deliberative council whose main remit is to establish standards and criteria for potentially polluting activities in the country. It is made up of representatives from all segments of society and is chaired by the Minister for the Environment.

The SNUC, Law 9.985 of 18 July 2000, establishes criteria and rules for the creation, implementation and management of conservation units. According to the law, conservation units are territorial spaces and their environmental resources, including jurisdictional waters, with relevant natural characteristics, legally instituted by the Government, with conservation objectives and defined limits, under a special administration regime, to which adequate protection guarantees are applied.

The system is made up of federal, state and municipal protected areas. In general, it aims to protect and restore biodiversity and promote sustainable development. It also brings the notion of environmental conservation together with valuing the culture of traditional populations, seeking to economically consolidate their activities.

It is managed by the following bodies, with their respective attributions: a consultative and deliberative body, CONAMA, with the task of monitoring the implementation of the system; a central body, the Ministry of the Environment, with the aim of coordinating the system; executing bodies, the Chico Mendes Institute for Biodiversity Conservation and state and municipal bodies, with the aim of implementing the SNUC.

The system establishes a series of parameters for the creation and management of protected areas, formulating categories that vary in terms of the degree of protection, ranging from units in which visitation is not even allowed to those that include industries and cities within them. Conservation units are divided into two groups with specific characteristics: integral protection units and sustainable use units. The first has the basic objective of preserving nature, with only indirect use of its natural resources permitted. The second aims to make nature conservation compatible with the sustainable use of part of its resources.

The integral protection group includes the following categories: ecological station, biological reserve, national park, natural monument and wildlife refuge. The main characteristics of these units are great scenic beauty, which is used for ecological and sporting tourism, and ecological exceptionalities, which are used for scientific research and environmental education. This category is clearly influenced by North American preservationism, which prohibits human habitation.

Of course, this does not mean that there are no inhabited units of this category, which are the result of the time when they were created without the participation of society and without adequate studies. The biggest adversity faced by these units is land regularisation. The land contained in these units must be in the public domain, except in the case of natural monuments, so the owners must be expropriated and duly compensated. However, the lack of resources and the slowness of this process are causes of conflict.

The sustainable use group includes the following categories: environmental protection area, area of relevant ecological interest, national forest, extractive reserve, fauna reserve, sustainable development reserve and private natural heritage reserve. These units are aimed at the sustainable use of natural resources and may be public or private domain, depending on their category.

Multiple economic and social activities can take place in these units. This is because they are often created as a way of organising activities that have a high impact on the environment. In this way, a legally-based management apparatus is created that allows the expansion of these activities to be restricted or at least regulated. Another characteristic of this group is to ensure that traditional activities continue to be practised in their territories. The way of life of populations who derive their livelihoods from the biodiversity of ecosystems is seen as compatible with conservation. Therefore, land-use planning and the promotion of traditional activities are one of the main objectives in these protected areas.Thus, conservation units are instruments for land management and land-use planning. They must be managed by an advisory council and, in some cases, a deliberative council, made up of the executing body and representatives of the social actors present in the territory. Of course, their creation involves various political interests, discussed both locally and nationally. Today, this process aims to be as democratic as possible, trying to involve society in the discussions. However, its unilateral creation generated conflicts that continue to this day.

2.4 Traditional Populations in Conservation Units

The implementation of the SNUC generates conflicts. The state's intention is to control the use of the territory in order to protect biodiversity and natural resources. Today, scientific studies and public consultations precede the creation of a conservation unit. However, the ways in which the territory is used, appropriated and controlled are diverse. Legal control of human activities in the unit takes precedence over other appropriations and uses of the territory.

Generally, sectors linked to productive capital, such as industry and agriculture, act against the implementation of protected areas because they limit their exploitation of resources. Similarly, traditional populations also generate tensions, as they have historically had no say in the demarcation and implementation of these areas (GUERRA and COELHO, 2009).

Today, the populations affected by the creation of protected areas are indigenous peoples, artisanal fishermen, country folk, caiçaras, holidaymakers with second homes, traders in general, wage earners, extractivists, researchers and tourists (VIANNA, 2008). The representation of nature conceived by environmental policies is imposed on local populations because it is a cultural approach transformed into a scientific concept. In this way, the importation of preservationist models of conservation units that are not compatible with the reality of the populations has led to the emergence of conflicts.

By imposing the vision of protected nature without human presence, many traditional populations were removed from their lands. The environment became synonymous with confiscation for them. In order to make their presence in protected areas compatible, categories were created for the use of traditional populations. In the national forest (Flona), the extractive reserve (Resex) and the sustainable development reserve (RDS) there is the promotion of the culture and activities of these populations, combined with the conservation of the local ecosystem. Some activities such as gathering, hunting, fishing and farming are permitted within them. Their cultural practices and identity are also legally protected.

In the case of Resex and RDS, the initiative to create the unit depends on the will of the traditional population. It is up to them to ask the executing agency to carry out feasibility studies for their creation. The land is then in the public domain, the concession of use is given to the communities and the management is exercised by a deliberative council. For this to happen, the population must be organised, live in a community and have a common territory linked to their cultural identity.

These units, then, are based on the idea of a traditional population to create an instrument that seeks to add the culture of communities whose activities are linked to the parsimonious use of natural resources to economic development. With the environment becoming a promising market, these populations have the potential to produce goods with added socio-environmental value.

The overlapping of different forms of appropriation/domination of territory is a constant in the process of creating and implementing protected areas. For this reason, the restriction of its use brings conflicts. However, the ideal of environmental conservation values the use and appropriation of territory by traditional populations due to the intimate relationship their activities have with local ecosystems and their cultural ties.

CHAPTER 3

Traditional Populations

3.1. The Idea of Traditional Populations

The emergence of social conflicts caused by the implementation of protected areas is the seed for the idea of traditional populations. Several protected areas had been created where local populations lived. In the reformist challenger's vision of conservation, man is synonymous with modern man, so any environmental area that was not part of the urban-industrial economy was considered a demographic vacuum. This is why regions such as the Amazon, considered uninhabited, attracted attention when they were historically occupied by rubber tappers, riverbank dwellers, artisanal fishermen and indigenous peoples (VIANNA, 2008).

The challenge, then, for the design of protected areas, was a less restrictive understanding of their management. Formally, at international level, this position was consolidated in 1994 with an IUCN proposal for protected natural areas recognising the positive role that native peoples play in conserving the environments they manage (Ibid).

Thus, quoting Clay (1985, apud VIANNA, 2008, page 209):

> Protected areas could guarantee the survival of habitats as well as native populations. Reserves could preserve traditional ways of life or slow down the pace of change to levels that are acceptable and controlled by local residents. Native populations could benefit from the protection of their rights over these areas or from the sale of products or income generated by tourism.

By realising that biodiversity and nature are not only part of the ecological environment, but also the fruit of a socio-cultural construction, natural resources are revalued by associating them with traditional knowledge. Therefore, by discussing the sustainable use of ecosystems and the right of local populations to reproduce their way of life, the idea of traditional populations is reinforced.

Various authors have discussed this idea. For Desmann (1989, apud Diegues 2000), they are considered "ecosystem populations" because they have a harmonious relationship with nature, due to their knowledge and sustainable traditional practices for managing natural resources, and are even considered producers of biodiversity. They are contrasted with what he calls "biosphere populations", which are linked to the global economy, high consumption and the power to transform nature. However, he considers the classification to be simplistic, since what exists is a continuum between one category and another.

From a Marxist perspective, traditional cultures are linked to pre-capitalist modes of production, typical of societies where labour has not become a commodity, where dependence on the market already exists, but is not total (Diegues, 2000). These societies develop ways of managing nature without aiming for profit. Their focus is on the social reproduction of the community and its cultural practices.

We know that no culture is static, and that the incorporation of cultural elements and the exchange of knowledge are part of the dynamics of society. Thus, the idea of traditional populations is controversial, since discussions about when these populations will continue to have these characteristics are pertinent. However, in the view of anti-modern conservationism, the presence of traditional populations within protected areas brings the idea that biological diversity and cultural diversity should be equally protected.

Therefore, in Brazil, a group of populations of artisanal fishermen, small subsistence farmers, caiçaras, caipiras, peasants, extractivists, pantaneiros, jangadeiros, riverine dwellers and indigenous people are considered traditional. They make direct use of natural resources through extractive activities and/or agriculture using technology that has a low impact on the environment. Their knowledge of the environment in which they live is passed down for generations through their social relationships. They have a strong bond with the territory they inhabit, as it is the source of their food and cultural practices. Their sense of belonging and identification with a specific group is closely linked to their territory (VIANNA, 2008).

These characteristics that make up the idea of traditional populations bring with them a trump card for conservationism and for social movements in the clashes over the right to territory. In a country of latifundia and land grabbing for speculation, such as Brazil, cases of land conflicts are latent, involving various groups. Therefore, the discussion around environmental conservation empowers these populations in the territories they occupy, showing that they are strong allies of ecosystems in potential destruction.

With the idea of traditional populations being discussed at international level, such as the International Union for Conservation of Nature and the Convention on Biological Diversity, the Brazilian government is now interested in developing policies aimed at this group. According to Vianna (2008, page 215), *"these discussions take place from the perspective of the possibility of populations occupying the territory of indirect use conservation units (...), in order to minimise conflicts, taking advantage of the ecological characteristics of these social groups for conservation.* *"*

The legal basis for the protection of these populations begins with the 1988 Federal Constitution, which establishes in articles 215 and 216 that:

> *Art. 215 - The State will guarantee everyone the full exercise of cultural rights and access to the sources of national culture, and will support and encourage the valorisation and dissemination of cultural manifestations.*
>
> *Art. 216 - Brazil's cultural heritage is made up of material and immaterial assets, taken individually or as a whole, which bear reference to the identities, actions and memories of the different groups that make up Brazilian society, including:*
> *I - forms of expression;*
> *II - ways of creating, making and living;*
> *III - scientific, artistic and technological creations;*
> *IV - works, objects, documents, buildings and other spaces used for artistic and cultural manifestations;*
> *V - urban complexes and sites of historical, landscape, artistic, archaeological, palaeontological, ecological and scientific value.*

Traditional populations therefore have a constitutionally guaranteed right to cultural territory, which is necessary for the exercise and development of the most diverse forms of knowledge, innovations and cultural practices such as music, tales, legends, dances, as well as craft techniques ranging from the management of natural resources to hunting and fishing methods and knowledge about ecological systems and species with pharmaceutical, food and agricultural properties (LOURIVAL, 2009).

In 1992, IBAMA created the National Centre for the Sustainable Development of Traditional Populations (CNPT), now part of ICMBio. One of its specific tasks is to promote the economic development of traditional populations based on sustainability, culture and the knowledge they have accumulated.

In 2007, Decree 6.040 established the National Policy for the Sustainable Development of Traditional Peoples and Communities - PNPCT. In this document, traditional peoples and communities are understood as:

> Culturally differentiated groups who recognise themselves as such, who have their own forms of social organisation, who occupy and use territories and natural resources as a condition for their cultural, social, religious, ancestral and economic reproduction, using knowledge, innovations and practices generated and transmitted by tradition.

The importance of territory for traditional populations is also reflected in the decree, adopting the term "traditional territory" as: "the *spaces necessary for the cultural, social and economic reproduction of traditional peoples and communities, whether they are used permanently or temporarily"*. This policy aims to promote sustainable development for these populations, as well as strengthening their territorial rights and valuing their identity.

The idea of traditional populations, then, arises to deal with existing conflicts over the appropriation of territories by communities that use them for their social and material reproduction. In addition to environmental preservationism, which has created areas to protect nature without considering their presence and activities, the fishing, mining and electricity industries and the agricultural sector are also examples of actors who dispute their territories and natural resources.

3.2. The Caiçaras

Fishing in Brazil

In order to understand the reality of the caiçaras, it is important to know about the development of fishing in the country, the policies aimed at this area and the relationship with the territory of different modes of production.

The Brazilian sea, with 8,500 kilometres of coastline and 4.5 million square kilometres of Exclusive Economic Zone, represents almost half of our entire land territory. Fishing activities, and not the occasional gathering of shellfish and other aquatic resources from the coast - which is an even earlier activity - are probably older than agriculture and livestock farming. Nevertheless, thousands of years later, fishing continues to play a fundamental social and economic role.

Fishing communities, maritime culture, maritime human communities, people of the sea, traditional fishing communities, coastal communities or, simply, peoples of the sea, represent in Brazil "a population contingent of approximately 800,000 fishermen and fisherwomen, involving two million people who produce around 55 per cent of national fishing production (CALLOU, 2010, 45).

In these communities there are cultural elements common to all of them, formed by the same influences that served to lay the primitive cultural foundations of coastal life (MUSSOLINE, 1972), transcending merely family boundaries to become a community activity, establishing a whole series of interactions between the residents, uniting them in co-operation. Within this social context, other aspects of culture, traditions and customs are also emphasised, giving the group formed a unique and concise identity. In contact with other contemporary cultures, these peoples hybridise, "developing particular *forms of* knowledge and social organisation for the use of *natural* resources *and the conservation of marine resources*"" (DIEGUES, 1993).

Most of the time, public policies have been designed to develop Brazil's fishing sector with these traditional cultures in mind. A historical analysis of policies aimed at fishing in Brazil shows that they have acted in two ways: to establish regulations and to provide incentives for production. The basis of these policies was the economic exploitation of natural resources, based on the modernisation of fishing activities.

Among the interventions carried out, two deserve to be highlighted: the Mission of the Cruiser José Bonifácio (1919-1924) and the Superintendence of Fisheries Development (Sudepe) (1962-1989). The first concerns the Navy's interventions in coastal fishing communities, with explicitly military interests, permeated by social and economic aspects. With regard to Sudepe, it can be seen that the guiding vision of its policies is based on the view that Brazilian fishing is "primitive and miserable" (Sudepe, 1963) and that investments will mainly be made in the industrial fishing sector through tax incentives.

From the 1960s onwards, fishing activity gained momentum. The policy to encourage fishing production began in 1967 with the enactment of Decree-Law 221. Until then, fishing in Brazil was predominantly artisanal and its production was basically geared towards the domestic market. Later, through this policy of tax incentives for fishing, so-called industrial fishing was developed, aimed primarily at the foreign market.

"The *results of these policies did not generate the desired development of either industrial or artisanal fishing* (PAIVA, 1967)." On the contrary:

> *modernisation and incentives for the industrialisation of fishing have led to the depredation of various species of fish and crustaceans, compromising the lives of coastal communities. In addition, fishermen and fisherwomen pointed to serious problems arising from overfishing, property speculation on the beaches and tourism, which has systematically expelled fishing communities from their traditional territories* (CALLOU, 2010).

In 1989, the Federal Government abolished the Superintendence in question and its attributions and competences were transferred to the Brazilian Institute for the Environment and Renewable Natural Resources (Ibama), of the Ministry for the Environment, Water Resources and the Legal Amazon. With the creation of the Special Secretariat for Aquaculture and Fisheries of the Presidency of the Republic (Seap/PR) in 2003, today

Ministry of Fisheries and Aquaculture (MPA), the government recognises the country's social debt to artisanal fishing and is drawing up a strategic plan for sustainable development.

In this way, the private conglomerates are fighting for their right to exploit and the traditional population for their historical and cultural rights to survive and territorialise. The dynamism of this process of fighting for space configures the contemporary dilemma between the dichotomy of the market economy and the preservation of traditional ways of life. Brazil's coastal zone is a space in the struggle between the ideologies of globalisation - *"which imagine themselves to be the bearers of the future, of pragmatism and reason* - and *ideologies of the foundation, the region, the root, the origin, the past, the origin and the feeling"* (ZAR\JR, 2000). Territory is therefore articulated in global, capitalist networks, which use it for material reproduction, and in local networks, articulated by traditionality.

Characteristics of Caiçara Communities

In times gone by, when the great blue expanse was something insurmountable for man, the coastline held the unknown. It populated people's imaginations with myths and legends about what existed after the waves broke, demonstrating that even without having been travelled by man, it was already symbolic territory. In Europe, at the time of the great navigations, the coastline represented the arrival of goods and people through the harbours, the threat of pirate or enemy invasion and the source of fishing resources and salt. In a society strongly permeated by Christian values, the sea was the dwelling place of creatures not desired by God. Therefore, only those who fished lived on the beaches (FREITAS, 2007)

When man became aware of the maritime territory, through the development of techniques, the dispute over it began, since society was in the midst of an expansion of markets and domination of the seas became a matter of national sovereignty for some states. In this way, society's relationship with the coast became more intimate, as its meaning changed with the change in the way it was appropriated.

Therefore, economically and culturally linked to the sea, caiçaras are *those communities formed by the mixture of the ethnic-cultural contribution of indigenous people, Portuguese colonisers and, to a lesser extent, African slaves. The caiçaras have a way of life based on itinerant agriculture, small-scale fishing, plant extraction and handicrafts"* (DIEGUES, 2000). This culture developed on the coasts of Rio de Janeiro, São Paulo, Paraná and Santa Catarina. It was formed between the great economic cycles of the colonial period, when export activities went into decline, thus reducing port activities.

The caiçaras have always had contact with port cities such as Paraty, Paranaguá and São Sebastião to sell flour, fish and handicrafts produced in their territories. Their goods are transported by land, along trails or paths, and by sea, using "vogue canoes". The tools they use on a daily basis for fishing and farming are collected in the forest and manufactured in the communities, buying what they can't collect in nearby towns. They make fishing nets, canoes, small boats, hoes and other tools.

For the fishermen's colonies, the coast is the place where they reproduce their way of life, in other words, the place where they carry out their economic and socio-cultural activities. The main activity of these communities is fishing, through which fishermen obtain their food and their products for sale. The territoriality of these people is closely linked to the fishing techniques they use. The fisherman who owns a trawler (a boat used to catch sardines) moves around in search of sardine shoals, so he has an extensive maritime territory where he knows the best spots to catch the fish (CARDOSO, 2001).

On the other hand, the fisherman who practises purse seining, a technique that consists of setting up a net at a certain point in the sea and only removing it a few days later, doesn't need to go to sea every day, so he has time to devote to another activity on the coast (farming, for example). Artisanal fishing consists of knowledge of the local physical environment (wind, fishing spots, tides, species' reproduction cycle) and social relations between fishermen and communities. According to Vianna, *"beaches are communal territories where social relations take shape"* (VIANNA, 2008). Taking the boats out of the water, meeting up early in the morning for another day of fishing, celebrations and information on events are all socialisations that take place on the beach.

In the passage quoted by CARDOSO, E. S (2001), Maldonado points out "that the *appropriation of the fisherman's territory is measured by the technological level of the fishing tools and, above all, by the maritime knowledge that each group builds and develops in its behaviour towards nature"*. The conservation of natural resources at sea, then, can be seen in the way fishermen relate to the sea, by rotating fishing spots, not catching young individuals and maintaining the quality of the waters.

As such, the coast is appropriated symbolically in terms of the communities' beliefs and social representations, and functionally in terms of providing fishing resources. Artisanal fishing is a collective activity, requiring a relationship of trust between fishermen and knowledge of the maritime territory.

The lands of the caiçaras are currently the object of desire by public authorities and environmental organisations. Because they live on beaches that are furthest from urban centres and have great natural beauty, caiçara fishermen have been expropriated from their land by real estate speculation, which sees a growing market for holiday homes in these areas. There is also pressure from environmental agencies, both governmental and non-governmental, to ensure that the last remaining patches of preserved Atlantic forest remain so, seeing the presence of caiçara populations as a potential risk of degradation.

Thus, Diegues (2000) emphasises the pressure that the caiçara territory has been under. suffering:

> *"Another process responsible for the disorganisation of caiçara culture is the fact that a large part of their territory has been transformed into protected natural areas. This transformation of their space for social and material reproduction into parks and nature reserves has resulted in serious limitations to their traditional activities of itinerant agriculture, hunting, fishing and extractivism, contributing to the emergence of conflicts with the administrators of these conservation units and to even greater migration to urban areas"* (DIEGUES, 2000: 76).

The caiçara populations that inhabit the states of Rio de Janeiro and São Paulo are

Paulo has been under strong pressure to occupy its territories. Both because of the high value of their land for the property sector and tourism, and because of the creation of protected areas to preserve remnants of the Atlantic rainforest.

CHAPTER 4

The Juatinga Ecological Reserve (REJ) - Paraty/RJ

4.1 . General Characteristics of the Municipality of Paraty

The municipality of Paraty belongs to the Costa Verde Region, the southern coastal region of the state of Rio de Janeiro, which also includes the municipalities of Angra dos Reis, Itaguaí and Mangaratiba. The total municipal area is 92,491.94 hectares (924.91 km2), which corresponds to 39% of the area of the Costa Verde Region and around 2% of the total area of the state of Rio de Janeiro.The Ilha Grande Bay region, where the municipality of Paraty is located, concentrates important forest fragments and marine environments in the state of Rio de Janeiro. The municipality of Paraty has the largest proportion of forested areas in this region, playing an important role in the connectivity of these forest fragments to the large fragments on the north coast of São Paulo. The municipality has more than 80 per cent of its forested areas, 70 per cent of which are highly conserved. The only other cover that is more than 2 per cent is grass (around 10 per cent). Built-up areas account for just 1.26% of the total area of the municipality of Parati, as shown in Table 1.

Table 1 - Distribution of Vegetation Classes and Land Use in the Municipality of Paraty/RJ.

Classes	Hectares	% of area
Rock outcrop	1.813,80	1,96
Forest in a medium or advanced stage of succession	65.344,49	70,65
Early successional forest	10.624,18	11,49
Shrub vegetation	1.717,08	1,86
Grasses	10.029,07	10,84
Restinga	457,02	0,49
Brejo	178,66	0,19
Mangrove	517,06	0,56
Beach	110,98	0,12
Rocky coast	251,16	0,27
Water	58,01	0,06
Built-up area	1.131,65	1,22
Wooded area	57,92	0,06
Agriculture	135,19	0,15
Exposed soil	51,53	0,06
Airport	14,14	0,02
Total	92.491,94	100

(Source: Participatory Master Plan of the Municipality of Paraty, 2010)

The vegetation that characterises the municipality is part of the Atlantic Forest biome, with a predominance of forest formations typical of this biome. These forests colonise a large part of the hillsides, especially the areas of steep slopes in the higher portions, inside the Serra da Bocaina National Park. Despite their small size in relation to the municipality, the mangroves and restingas that remain in the municipality are still important preserved areas of the coastal plains.

The conditions of the regional terrain, characterised by a small area of coastal lowlands and extensive areas of steep, high-altitude slopes, making access to large areas of the region difficult, have contributed decisively to the degree of conservation of the local vegetation. The municipality has only 9% of its area characterised by flat land and 27% by valley bottoms, while hilltops, mountains and slopes account for 65% of the entire municipal area.

Table 2 - Distribution of Topographic Positioning Classes in the Municipality of Paraty/RJ.

Topographic Positioning Class	Hectares	% of area
Mountain and hill tops	25.279,21	27,33
Slopes	34.198,31	36,97
Plan	8.354,89	9,03
Valley funds	24.659,18	26,66
Total	92.491,70	100

(Source: Participatory Master Plan of the Municipality of Paraty, 2010)

The history of the Paraty region shows the economic and social dynamics that result in its importance for conservation today. The economic cycles that have developed in the country have alternated between moments of apogee and decadence, dynamising local activities. The city, with its port vocation, was linked to the slave trade, the export of sugar cane, coffee, gold and fish to Rio de Janeiro and São Paulo. At one time, Paraty's harbour was the most important in the country in terms of exports.

Its colonisation dates back to the 16th and 17th centuries, when, with the granting of sesmarias, agricultural activities were set up in the country to predominantly serve the demands of the foreign market. Due to its geographical formations, the municipality served as a place to embark and disembark products, mainly sugar cane. As it was an active centre in the country's economy, it also attracted traders in rice, corn, beans, coffee, brandy and flour from the highlands. However, the sugar cycle was not as important in Rio de Janeiro as it was in the northeast and port activities began to decline as the price of the product fell (DIEGUES, 1994).

With the gold cycle in the mid-13th century, Paraty became a promising town, receiving the metal from Minas Gerais and the São Paulo plateau. The trails created by the Guaianás Indians that connected the entire region were used to transport this merchandise (VIANNA, 2008). Thus, the capitalist economy found an existing infrastructure, which contributed to local success. However, with coffee on the rise in the market, the attention of traders and the investments of the time began to turn to the interior of the country, relegating the coast to a secondary role, linked to port and subsistence activities.

Subsistence farming on the coast was intended to supply the monoculture farms and provide the population with food. The population centres were also linked to the larger centres through trade in manioc, fish and cachaça. These were two economic circuits: subsistence farming, maintaining a small local trade, and the export of what was produced in Brazil (Participatory Master Plan of the Municipality of Paraty, 2010).

In the 19th century, the port of Santos and the Dom Pedro II railway were inaugurated. The flow of goods turned to other ports, and so Paraty lost its reputation as an export town. At this time, the town suffered a major population decline, also attributable to the abolition of slavery.

Going into the 20th century, Paraty was still growing sugar cane for the production of brandy. The construction of several roads linking the cities of northern São Paulo with the coast expanded the integration of the region, but the connection between the coastal towns and villages was still via trails or the sea. Its physical characteristics, full of coves and rocky shores, made access to the towns very difficult, and "vogue canoes" were widely used to transport people and goods. This isolation of the municipality contributed to strengthening the way of life of the communities that lived in the region, the caiçaras (Ibid).

An important event that changed the dynamics in the municipality of Paraty and the region was the construction of the BR-101 motorway linking Rio to Santos. As a result, tourist interest in its beautiful beaches intensified and attracted new residents. The tourism industry took hold, building allotments, landfilling, draining and degrading the native vegetation. In a short space of time, tourism became the basis of the local economy and is now the main economic activity. The pressure on the properties of long-time residents is now great. Their land is coveted by businessmen in the area, and they are easily persuaded to sell because of the tempting prices offered. As a result, tourism has become the new conditioning factor for life in the municipality, with most of the shops and services being geared towards this activity (Ibid).

Particularly for the municipality of Paraty, the SNUC is a very important environmental management tool. The physical and biotic characteristics of this region present peculiar conditions for the conservation of both its biodiversity and its landscapes. The importance of the region's environmental characteristics is reflected in the large number of conservation units that overlap Paraty's territory - around 80 per cent of the municipality's continental area is occupied by Conservation Units, with many of the areas overlapping each other (Ibid).

Currently, the municipality has a total of 5 (five) Conservation Units in its territory, 3 (three) of which are federal: the Serra da Bocaina National Park, with 51.8% of the municipal mainland area; the Cairuçu Environmental Protection Area, with 18%; and the Tamoios Ecological Station with around 0.1% of the area of the municipality of Paraty; 2 (two) state: the Juatinga Ecological Reserve with 10.6%; and the Paraty - Mirim State Leisure Park with 1.3%. (Annex 1 - Map of PAs in the Municipality of Paraty) It should be emphasised that the data presented does not represent the total areas of the conservation units, but only the areas of these present in the municipality and discounting areas of overlap with other units. For example, the Cairuçu APA, despite being completely within the territory of Paraty, with its 33,800 hectares, only 17,141 hectares are shown in Table 3. The APA is complemented by the sum of the areas of the Juatinga Ecological Reserve and the Paraty - Mirim State Leisure Park (Participatory Master Plan of the Municipality of Paraty, 2010).

Table 3 - Distribution of Conservation Units in the Municipality of Paraty/RJ.

Category in the SNUC	Name of Conservation Unit	Hectares	%	Managing Institution
Sustainable Use	APA Cairuçu	17.141,93	18,5	ICMBio
Comprehensive Protection	Juatinga Ecological Reserve	9.796,98	10,6	INEA
Comprehensive Protection	Serra da Bocaina National Park	47.870,48	51,8	ICMBio
Comprehensive Protection	Tamoios Ecological Station	99,15	0,1	ICMBio
Uncategorized on SNUC	Paraty - Mirim State Leisure Park	1.191,54	1,3	INEA
Total area protected by PAs		76.100,08	82,3	
Total municipal area		92.491,94	100	
Total area not protected by PAs		16.391,86	17,7	

(Source: Participatory Master Plan of the Municipality of Paraty, 2010)

4.2 - General Description of the Reserve and the Residents

The Juatinga Ecological Reserve is a peninsula located to the south of the municipality of Paraty, between the tip of Escalvado and the tip of Trindade, in the Bay of Ilha Grande (see: Map of PAs in the Municipality of Paraty). The area, which has great biological diversity, is made up of rocky coastlines, steep cliffs and remnants of the Atlantic Rainforest, associated with secondary forests, sandbanks and mangroves.

According to State Decree no. 17.981 of 30 October 1992, which *"creates the Juatinga Ecological Reserve in the municipality of Paraty, and makes other provisions"*, its boundaries are on one side by the Saco do Mamanguá, on the other and frontally, by the open sea and, at the back, by an imaginary straight line which, starting from the point known as Cachoeira do Cocal (on the Canto Bravo side of Praia do Sono), reaches the place known as Porto do Sono (at the end of Saco do Mamanguá).

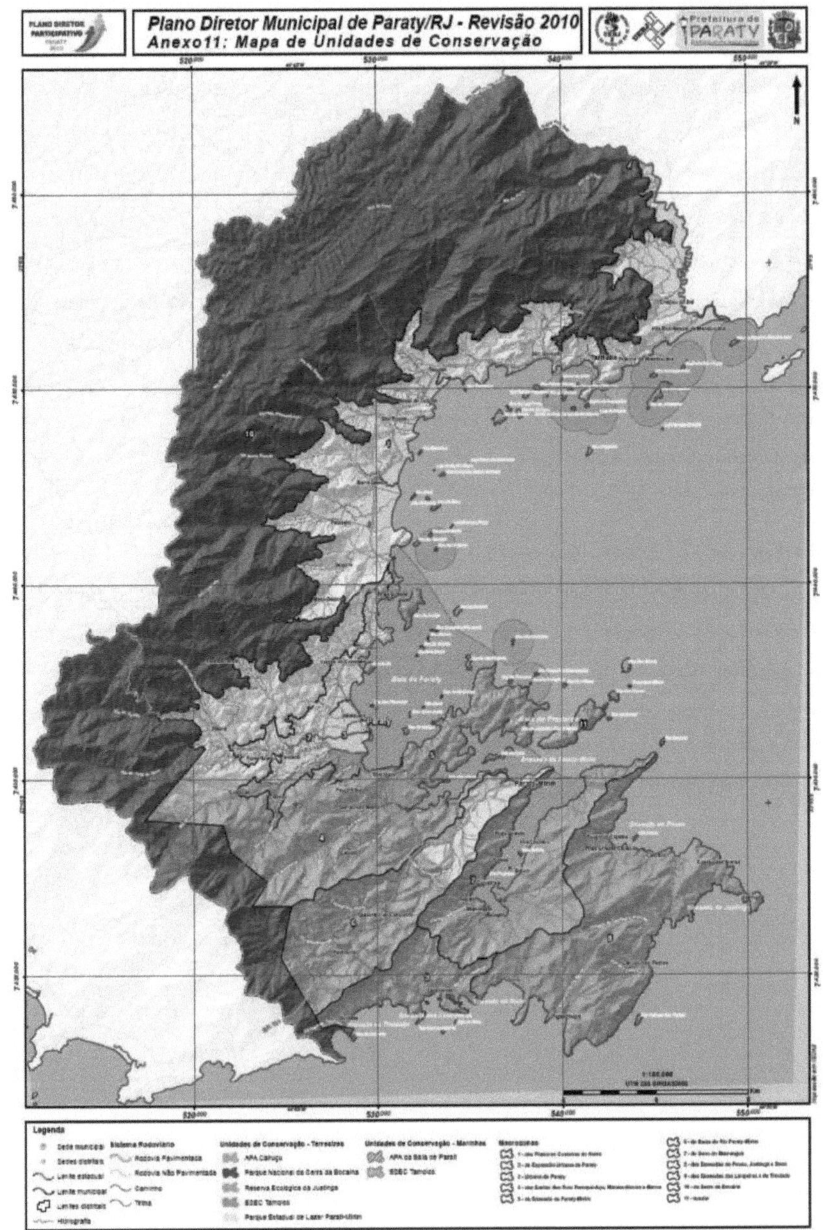

Map 1 - PAs in the municipality of Paraty - RJ. (Source: Participatory Master Plan of the Municipality of Paraty, 2010)

The geographical characteristics of the Juatinga peninsula make it difficult to access. The sea there is rough and the connection between the beaches is via trails. It has 13 settlements with an exclusively caiçara population. These settlements are characterised as the place of life of a group with a specific identity. They are identified by reference to the name of the beach or geographical features such as rocks and peaks. The inhabitants are linked to each other through relationships of compadrio, vicinities or kinship. Juatinga's settlements resemble rural neighbourhoods, considered to be the minimum sociability units of caiçara life (VIANNA, 2008).

On the north coast, from Saco do Mamanguá to the tip of Juatinga, lies Cajaíba Bay, where there are seven beaches. The only uninhabited beach is Praia Deserta, while the others (Praia Grande, Itaoca, Galhetas, Escaléu, Ipanema and Pouso da Cajaíba) have caiçara settlements. The population of Saco do Mamanguá is scattered along its 9 kilometres, with denser settlements on the beaches of Cruzeiro, Baixio and Ponta da Romana (DIEGUES, 1994). On the south coast of the peninsula, from the tip of Juatinga to Praia do Sono, people live on the beaches of Martim de Sá, Ponta Negra and Praia do Sono, as well as on the Cairuçu das Pedras and Saco das Enxovas coasts (VIANNA, 2008).

The territoriality of the caiçara populations in the reserve is expressed in the configuration of space. The houses, generally far apart, are grouped together in family clusters, arranged along paths or trails. The house, like the backyard, are private spaces. However, backyards are not demarcated or fenced off and are often used as a place for everyone to pass through. Even so, they are an important productive unit, as they grow spices, fruit trees, medicinal herbs and wood. *"It is the place of passage, the border between the house, a private place, and the forest, a common place."* (DE FRANCESCO, 2010). The backyard is home to the flour mill and where small animals are kept.

Photo 1 - Caiçara backyard, Saco do Mamanguá (Source: Lucas Grisolia, 2011).

Diegues, 1994, presents caiçara activities and their material and symbolic representations in the territory. He presents the "roças", a place where itinerant farming takes place, which are made in places far from the houses. When cultivated, the garden belongs to whoever is working on it. A month before planting, the land is cleared by burning. They mainly grow manioc, bananas, green maize, sugar cane and beans. As these are remote areas, sometimes hours away, the caiçaras can stay for days, working and living in temporary thatched ranches. Normally, several species of manioc are planted in order to reduce the crop's vulnerability to pests and diseases (DIEGUES, 1994, page 74). This cultivation strategy shows how this community aims to maintain biodiversity in order to reproduce their way of life.

The areas used for hunting, the forests, represent the working space of this population, where the search for food takes place. These change according to the abundance of fauna. Normally this activity is carried out during the day, with the hunter leaving in the morning and returning in the afternoon. However, the abundance of species has been decreasing due to the movement brought to the region by tourism and the predatory activities of extracting palm hearts and other fruit trees that provide food for the fauna (Ibid).

Plant species are extracted mainly for food, to build houses, canoes, boats, fishing gear and various types of handicrafts. Traditional culture is revealed not only in the precise knowledge of the species, but also in the respect for the phases of the moon when extracting wood, preventing them from being attacked by termites or splitting. There is also the selection of species to be harvested, with the aim of not depleting forest resources. The forest, then, is a place to search for the resources needed to maintain caiçara life (Ibid).

Artisanal fishing is one of the most important activities for the residents of Juatinga, as it represents the basis of the local community's diet and main source of income. The fishermen's empirical knowledge determines the best locations and fishing methods for the desired species. With a strong influence from indigenous and Portuguese methods, the caiçaras use trammel nets, first made of cotton and now of nylon, to fish for parati (Mugil curema) and mullet (Mugil platunus), which are abundant in the region. Fishing grounds, or enclosures, are clusters of wooden logs placed in strategic locations to attract marine organisms looking for food and protection. Fishing is then done with a handline. Another method also used is the seine net, which doesn't require the presence of the fisherman every day, leaving him free for other activities (Ibid).

Trawling, which consists of a bag-shaped net pulled by boats, is currently illegal. This activity had been carried out by fishermen from outside the reserve in search of white shrimp (Peneaus schimitti), which are abundant in Saco do Mamanguá. This type of fishing had a major impact on the biodiversity of the area, resulting in its condemnation by local residents and its legal prohibition.

The main boats used for both transport and fishing are canoes. Made from wood extracted from the forest, such as guapuruvú or cedar, they are indigenous manufacturing techniques. The whaleboat is another type of boat found in the reserve. Between 8 and 12 metres long, it was originally built by the caiçaras of Santa Catarina, but was adopted in the region.

Photo 3 - Two lifeboats, Paraty - Mirim (Source: Lucas Grisolia, 2011).

The mobility of the caiçara in their territory is closely related to marriage, their economic activities, their sociability and, more recently, the land issue. When the young caiçara wants to get married, he builds his house with his family, or with the family of his future wife, which can be in other villages or centres. They also travel along trails and paths between the centres to carry out activities such as fishing, farming and buying and selling goods. Events such as festivals, football matches and visits to relatives also drive the circulation of people and information in the territory. Land ownership is also another factor in changes in the place where people live. The strong pressure from property speculation causes caiçaras to move to other settlements (VIANNA, 2008).The activities presented by Diegues and Vianna are reflected in the region's landscape. The territoriality of the caiçaras can be seen by analysing their way of life and noting how the elements of their activities are arranged in the territory, showing that it is being used materially. Their symbolic representations are present in their religiosity, tales and myths, which also influence the appropriation of the territory.

4.3. Conflicts in the Reserve

The process of occupation of the southern region of Rio de Janeiro has given the municipality of Paraty the presence of traditional populations. Caiçaras, quilombolas and indigenous people of various ethnicities use the territory with traditional practices, overriding other forms of land appropriation, such as property speculation, disorderly tourism and conservation units.

47

Thus, driven by the strong pressure that these populations have been under to leave their territories and by the institution of the National Policy for Traditional Peoples and Communities, which gave them legal backing, the Forum of Traditional Peoples and Communities of the Southern Region of the State of Rio de Janeiro and Northern São Paulo was created in 2007. The Forum, as an independent organisation with no legal personality, brings together representatives of the caiçaras, quilombolas and indigenous peoples, forming a joint position on the conflicts.

According to the document "Consultancy to instruct procedures relating to the characterisation and resolution of conflicts", developed for the Ministry of the Environment under the coordination of the Secretariat for Extractivism and Rural Development/Coordination of Agroextractivism in 2009, of the 20 communities that currently take part in the Forum, 18 are in protected areas. Their main conflicts are related to the use and ownership of their territories, linked to the overlapping of territories with protected areas, property speculation and uncontrolled tourism.

The Juatinga Ecological Reserve, managed by the Rio de Janeiro state environmental agency, INEA, formerly the State Forestry Institute, is considered an integral protection conservation unit, i.e. it only allows indirect use of natural resources. State Decree no. 17.981 of 30 October 1992, the normative act for its creation, states in articles 1, 3 and 4 that:

> Art. 1 - The Juatinga Ecological Reserve, of a non-building nature, is hereby created in the Municipality of Paraty (...).
>
> Art. 3 - The State Secretariat for the Environment and Special Projects (...) will issue the necessary instructions for the effective implementation of the Juatinga Ecological Reserve, in compliance with the environmental legislation in force.
>
> Art. 4 - The State Forestry Institute Foundation (I.E.F./RJ) will develop a specific Environmental Education programme, with the aim of fostering local caiçara culture, making the use of natural resources compatible with conservation precepts.

The inconsistency lies in considering the presence of the caiçaras, specifying that the state environmental agency must develop programmes to promote their culture, and characterising the reserve as non-building. As a result, it will determine that construction is forbidden inside the reserve, with only the remodelling of existing buildings being allowed when duly authorised by the management body. This imposition is in direct conflict with the caiçara way of life, because in their family relationships, the son who marries must build his new home, usually close to his relatives, and raise his family (VIANNA, 2008). The decree thus prevents the expansion of caiçara society in their territory.

Article 3 of Law No. 1,859 of 1 October 1991, which authorised the creation of the reserve, specifies that the Executive Branch, i.e. INEA, should be responsible for regularising the REJ's land ownership. However, almost twenty years after its creation, no measures have been taken to fulfil the provisions of the law, leaving the caiçara population without legal support for their possessions.

Another issue is the reserve's overlap with the Cairuçu APA in its entirety. This conservation unit was created before the Juatinga Ecological Reserve and one of its objectives is to make the caiçara way of life compatible with the conservation of the region's ecosystems. In the APA's management plan, the area of the reserve was zoned as a Wildlife Preservation Zone, a Coastal Conservation Zone, with the possibility of management and cultivation by the caiçara communities provided it is done on a sustainable basis, and the Caiçara Village Zones, delimiting the perimeter that can be occupied exclusively by the caiçara community. These are zones designed to fulfil the rights of these communities in terms of expanding the homes of their children and grandchildren and preventing occupation by outsiders, such as holidaymakers and speculators. However, as well as not recognising this zoning, INEA has still not drawn up a management plan for the unit, which would be essential for its effective implementation.

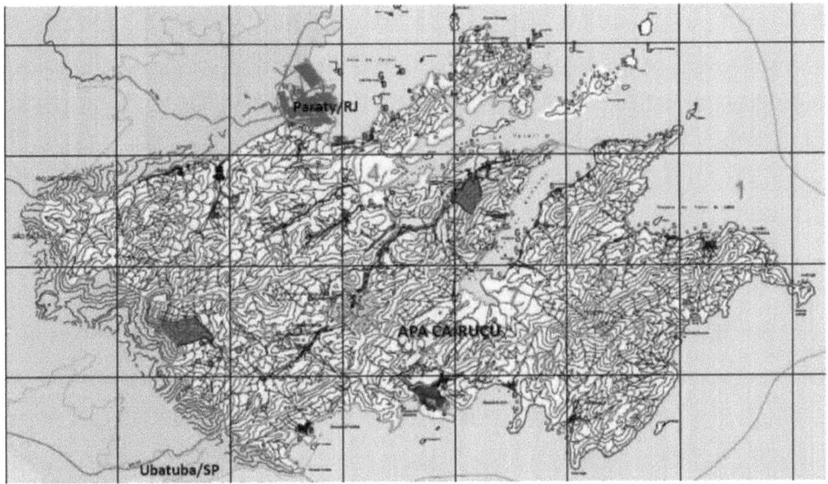

Map 2 - Boundaries and zoning of the Strategic Areas of APA Cairuçu, 2005 (Source: Abirached. C, 2011: 52).

49

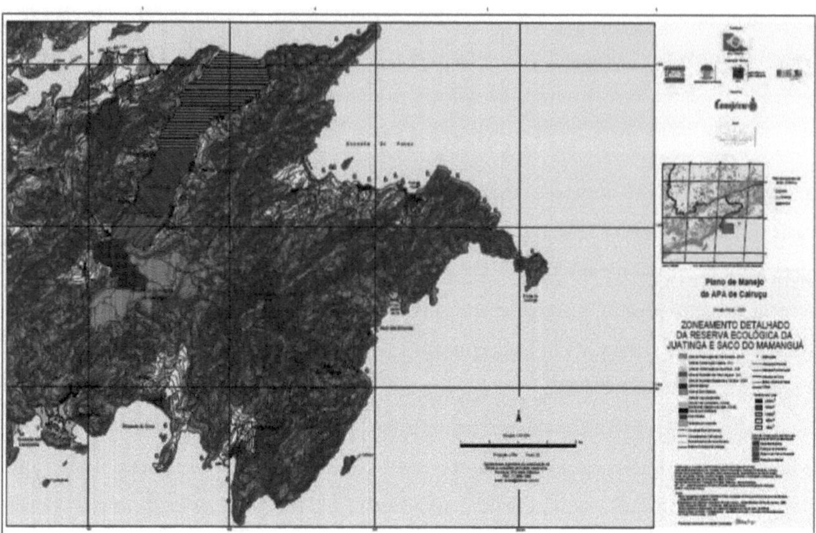

Map 3 - Detailed zoning of APA Cairuçu (Source: APA Cairuçu Management Plan, 2004)

The main conflicts faced by residents of the reserve in relation to the management of the conservation unit are the repression of traditional communities by the management body, coupled with the difficulty of access and dialogue with it. There is still bureaucracy and delays in issuing the authorisations requested, as well as the different treatment of the caiçaras communities in relation to other occupants of the reserve, such as holidaymakers. With regard to the use of natural resources, the suppression of native vegetation (vines, palm trees or wood) for the construction of houses, canoes or other caiçara infrastructure is prohibited without prior authorisation from the management body. The cultivation of subsistence crops and hunting are also prohibited or controlled.

Another problem is the lack of public policies. The fact that the area is an indirect use reserve is used as an argument for restricting the supply of electricity and basic sanitation, infrastructure (transport and paving), rubbish collection, basic education and health. This is one of the factors in the exodus of the caiçara population.

Property speculation takes place through the illegal buying and selling of land by land grabbers or developers, who expel communities from their original lands. Even though it is forbidden, there are large holiday homes inside the reserve that privatise beaches and restrict caiçaras' access to places they used to go to frequently. This logic of private appropriation runs counter to the caiçaras' communal way of life, in which the beach is a common place for all, freely accessible and a space for socialising.

Photo 4 - Holiday home surrounding the beach, Saco do Mamanguá (Source: Lucas Grisolia, 2011).

Other points of note are the existence of projects that have already been submitted to the environmental agency for bidding, and cases of violence as a way of intimidating and expelling the caiçaras from their land. On Martins de Sá beach, there is a legal dispute between a family that claims to own the land and wants to build a resort in the area, and the caiçaras who have occupied it for generations. At Praia Grande da Cajaíba, caiçara families have suffered physical and psychological violence from the family of the land grabber Gibrail Nubile Tannus, who is known in the region for taking over most of the land occupied by caiçara within the reserve (LOURIVAL, 2009).

The above-mentioned document also mentions INEA's destruction of the thatched ranches built by caiçaras on Praia Grande da Cajaíba, where their boats and fishing tools were kept. On the other hand, holiday homes and heliports continue to exist on other beaches, showing the agency's different treatment of the reserve's occupants.

There is also an issue related to the legal attributions of Paraty City Hall. In its preliminary draft of the Master Plan, there is no mention of the demarcation of caiçara territory, which goes against Resolution 34/05 of the National Council of Cities in its article 5º , where "The *establishment of Special Zones, considering the local interest, should: II - demarcate the territories occupied by traditional communities, such as indigenous, quilombola, riverine and extractive communities, in order to guarantee the protection of their rights"* (Ibid).

Fruit of the era when conservation units were created unilaterally, the Juatinga Ecological Reserve has latent conflicts between social actors who show different interests in relation to the use of the territory. Entrepreneurs show that their interests are focused on tourism and holiday occupation within the reserve. For Paraty City Hall, it's in the interest of encouraging developments that can bring more capital into the city. Meanwhile, the side with less decision-making power, the caiçaras, being on the margins of the current economic system, are trying to articulate themselves politically to defend their interests. Within these relationships, the public environmental agency, INEA, which manages the territory destined for the conservation of the Atlantic forest and caiçara culture, is inert and complicit in the actions of private interests. It is important to note that the reserve is in the process of being re-categorised to fit into a SNUC category. In this way, an attempt is being made to better meet the demands of the caiçara populations and minimise the pressures they are under. However, it is necessary that the managing body is really managing for the traditional communities and for the conservation of the ecosystem.

CHAPTER 5

The Life of the Caiçaras of the Juatinga Ecological Reserve

In order to better understand the relationships between the actors present in the Juatinga Ecological Reserve and how these reflect on the daily lives of its residents, an exploratory investigation was carried out. It took place from 3 to 7 April 2011 in the municipality of Paraty/RJ, with the aim of collecting statements from people who work directly on environmental issues in the region, managers of the municipality's conservation units and residents of the reserve. The data collected in the exploratory investigation consisted of statements from environmental analyst Carlos Felipe Abirached, who developed his master's thesis in the region; Rodrigo Rocha, head of the Juatinga Ecological Reserve; and Eduardo Godoy, head of APA Cairuçu. Statements were also taken from the caiçaras who live in Saco do Mamanguá.

5.1.   Testimonials

Testimonial I - Carlos Felipe Abirached: ICMBio Environmental Analyst.

"The *history of state management is bad. The reserve was created with the aim of conserving biodiversity, but it was also created to promote caiçara culture, as stated in the creation decree. In the law that authorised the creation of the Juatinga ecological reserve, the Rio de Janeiro state government was required to regularise the caiçaras' possessions, which has not been done to date. That was the legal directive, but INEA, which has a preservationist stance, considers the Juatinga Ecological Reserve to be fully protected, but if it was created to support the caiçaras, it could never have been incorporated as fully protected in the agency's institutional vision. This is a serious mistake.*

From this perspective, IEF, now INEA, restricted the activities that characterise the caiçaras. Farming was simply made impossible. And the caiçara is a farmer and has always supplemented his diet with fish. Today it's the other way round, because when the conservation units were created, farming was banned because it involves cutting down, burning, planting, abandoning and then moving to another area on a rotation basis. By making this impossible, you're already taking away a little of the caiçara culture. Renovations, extensions and the construction of new houses have always been opposed on the pretext that, based on the decree, they are non-building areas of the REJ. However, this should be aimed at non-Caçarans, especially holidaymakers who have a fat eye on the Caçaran areas of the Juatinga Ecological Reserve. The environmental agency comes up with the strategy of not being able to renovate, not being able to build, not being able to change the roof on your house, not being able to farm and not being able to hunt. This has created a cycle of private appropriation of the caiçaras' land, as they are unable to maintain their traditional practices, which are forbidden by environmental authoritarianism, and are left to work with tourism.

I ask: did the caiçara sell his land?

"No, either he ended up offering his land, or he received an offer to buy it (the house or the land). If someone from outside buys land there, the caiçara's son or grandson won't have anywhere to live. The areas there are very tight, they're between a short coastal plain, then there's the slope of the hill on one side and the sea on the other. So either they occupy an area susceptible to geological risks (landslides), or they live here in the city in neighbourhoods where many caiçara people from the whole region have been driven out by the real estate front.

In practice, the management carried out by the INF (now INnA) was very unfavourable to the caiçara. There were inexplicable restrictions, all under the influence of a view that the area was to be fully protected. For the caiçara, this implies diverse uses that reveal their way of life! It's worth remembering that the RnJ overlaps with the APA Cairuçu, created ten years earlier. The APA was created to be a buffer zone for the Bocaina National Park, to conserve biodiversity and also to support the caiçara who are integrated into this ecosystem. nssa APA has created appropriate zoning, which INnA doesn't seem to take into account.

The APA has an advisory council and zoning. The caiçara villages are treated in a special way in the APA's zoning and have become a zone for exclusively caiçara use and occupation. Outsiders can neither build nor renovate houses in these zones. It's a zone for the expansion of caiçara families, their children and grandchildren. But this hasn't been happening for a number of reasons. A lot of holidaymakers are appropriating caiçaras' houses or land. Then the question arises: but how do they manage to build? Many of the constructions that can be seen in Saco do Mamanguá, for example, as well as throughout the REJ, could not be authorised, because after 1992 that area is non-buildable for non caiçaras according to the correct legal interpretation of the case, since the reserve was created to support the caiçaras.

So the APA Cairuçu zoning provided for this. The Caiçaras Village Zone is exclusively for caiçaras. At the top, in the hillside areas, is the Coastal Conservation Zone, where caiçaras can carry out agricultural management, timber and non-timber forestry, and extractivism. In the upper areas are the Wildlife Preservation Zones. It's a full protection zone regime, except that if the caiçara needs to fell a stick, pick up the wood, work on the canoe and drag it down to finish it off, he can ask ICMBio, because the APA's zoning allows it. The APA Cairuçu zoning met the need for expansion of the caiçara villages; it met the management areas, including the swidden; and it met the demand for timber. It is a zoning system that is very well suited to the caiçara way of life. APA Cairuçu's zoning is what should in fact govern the areas where caiçaras live and use their land.

INEA's perspective is as follows: to re-categorise the Juatinga Ecological Reserve to fit in with the SNUC. The majority view of the agency seems to be to fulfil the goals of full protection. So the REJ area would become a state park in the higher areas, perhaps from level 100 or 200 upwards. In the lower areas, where the caiçaras are, the idea will probably be to de-allocate the REJ area and make it the APA Cairuçu, or create an RDS, which would be better.

We need to change the course of history. You can't simply restrict all the practices of the caiçaras, otherwise you'll exhaust all the possibilities for the caiçaras to maintain themselves as such. The predominant strategy is to undermine all the caiçara forces so that they agree with the speculator, leave *their* land and move to the *outskirts of the city, leaving their land free for resorts, condominiums, developments or a state park. As a result, an entire caiçara culture is lost.*

Added to this is the issue of revising Paraty's Master Plan. Municipal land use planning is the responsibility of the municipality. But the City Council is using the pretext that its competence to organise the territory is absolute. And it's not. So much so that the federal government and the states also have the power to organise the land through the creation of protected areas and also to organise these areas for protection. The City Council seems to want to dismantle the legal regime of the APA Cairuçu zoning so that houses, condominiums and developments can be built in the caiçaras' villages. To do this, it claims that its competence is superior to that of the state and the Union to organise the territory.

Since 1992, when the REJ was created, all the houses and mansions in Saco do Mamanguá have been illegal. Do they fulfil their social function? Today in Saco do Mamanguá it seems that most of the caiçaras don't want the holidaymakers to leave because they live off tourism, many earn half or one minimum wage as caretakers, selling the most expensive shrimp to tourists. So the eventual expropriation of at least part of the property for a possible RDS in the area will have to address these issues.

So my vision is to work on valuing caiçara culture, developing projects, securing the territory, bringing alternative income, working with young people. Young caiçara people today need an alternative. They can be environmental monitors, lead tours along the trails, take tourists to see the caiçara way of life, but they need to be worked on, they need to be trained in their skills. Sometimes they even need to speak another language, but that's not why they'll stop being caiçara. We have to give the caiçara the chance to choose their own destiny, and not have an environmental agency, a land grabber or the City Hall unilaterally rule over a territory that has been occupied by the caiçara for centuries."

Testimonial II - Rodrigo Rocha: Head of the Juatinga Ecological Reserve

I ask if the future of the region has already been mapped out.

"A number of socio-environmental characterisation studies are now being carried out on the reserve as part of the re-categorisation process. These studies are being carried out by a consultancy firm from the region, hired by INEA.

So I'll give you an overview of the reserve. The reserve was created in 1992, so it's 18 years old, but to this day it doesn't have a management plan, it doesn't have land regularisation, the level of implementation of the reserve is very low. Although it has an administration, with an appointed administrator, a small technical team, we have a car and a headquarters structure, but the level of implementation of the unit in the field is very low, since we haven't made any progress with land regularisation, the management plan, the issue of public use planning, these are things that are still in their infancy.

And this is the responsibility of INEA, which is the unit's current management body. It is currently a body that has a wide range of competences, including the management of conservation units, both full protection and sustainable use. So, ten years after the SNUC, the reserve is belatedly entering this re-categorisation process, since the SNUC defines that this re-categorisation should take place within the first two years after the law comes into force. In fact, it's important to say that what is being done is just a study, it's just a subsidy for the re-categorisation process, the study itself doesn't mean re-categorisation, it just points the way."

Question: the REJ is within the APA Cairuçu and the APA has management plans. Does the REJ recognise this management plan?

"No, the state's environmental administration doesn't recognise the APA Cairuçu management plan as a plan that applies to the Juatinga Ecological Reserve, although we often use the management plan at our base here, we base our opinions on articles from the APA Cairuçu management plan.

Not least because of the lack of regulation due to the reserve not having its own management plan. For example, when we receive requests for buildings within the ZPVS (Wildlife Preservation Zone) of APA Cairuçu, we base them on this. There can't be any kind of construction in this area. When we receive a request to remove a fallen tree, to make a canoe, or to extract a non-timber forest product, we even authorise it, we have nothing to oppose. Bearing in mind the existence of the caiçara communities that *depend on these resources'*.

Why doesn't the REJ consider the APA Cairuçu management plan?

"This is more of an understanding on the part of the central administration. At the time the APA Cairuçu management plan was drawn up, sponsored by the NGO SOS Mata Atlântica, there was an attempt to liaise, to promote an understanding so that the document that was being drawn up would in fact be applied to both conservation units. There is even a detailed zoning map for the REJ overlap area in the APA Cairuçu management plan. However, the environmental administration, at the time the former IEF, never accepted this document as a reference for the territorial management of the REJ. Even today, it doesn't have it as a reference document. If you ask me why, perhaps one of the reasons is because of the areas that were established as a caiçara village expansion zone. Perhaps there was no understanding of this zoning. Now all of this is going through a review process, because as soon as this re-categorisation process begins, it has implications not only for the REJ, but also for the APA Cairuçu. To the extent that re-delimitation is carried out, it could affect the zoning provided for in the APA's management plan."

I ask: is the management of the REJ articulated with the management of APA Cairuçu?

"Today it's totally articulated. Today, fortunately, REJ and APA do practically all the procedures together. All the inspections, surveys and community meetings we always try to do together. We have projects in partnership for the region. We're also very integrated in the study for re-categorisation."

I ask: how is REJ's relationship with the caiçaras?

"The relationship today is conflictual. Why is that? When it was created in 1992, the legislator sought to create some elements in the law to protect caiçara culture, and so the law also came out with this important attribute. Making caiçara culture compatible with conservationist precepts. However, over the years, for the state's environmental administration, this has taken on a different interpretation. An interpretation that is much more restrictive than fostering caiçara culture. Added to this is the change in the region's economic cycle, in which some traditional activities have gradually been replaced by other activities, such as tourism.

Today, for example, fishing remains an important activity, perhaps the main economic activity, more so than tourism. However, agriculture, which used to be the second main activity, no longer is. These caiçara gardens have gradually been abandoned, somewhat due to pressure from the state, because the traditional practice is fallow farming. You cultivate for a certain period of time, two or three years, and when the soil starts to deplete, you leave the land to rest and move on to cultivate a new area. This began to cause friction, because when he left that area, it became a capoeira and then, at a more advanced stage of forest cover, a dirty capoeira. As a result, this area fell into a situation of environmental restriction. The Forest Code itself and the Atlantic Forest Law place these restrictions. So over the last few years this agricultural activity has been replaced by other complementary activities, mainly activities linked to tourism.

We have very recent data showing that within the REJ area there has been an increase in forest cover. So areas of pasture and low shrub vegetation have become vegetation at an early stage of regeneration and those areas that were at an early stage are now at a medium stage of regeneration. In general, forest cover has increased. The main reason for this was the abandonment of agricultural activities. Fishing continues to be a very strong activity, but agriculture has been significantly reduced"[1] .

I ask: in relation to the mobility of the caiçara within the REJ, can they expand within its territory?

"Today the state restricts construction."

I ask: even for the caiçaras?

"Different managers have dealt with this issue in different ways over the 18 years of the REJ. The current INEA administration believes that in order to solve this problem, re-categorisation will provide a definitive way forward. Provisionally, the current administration authorises renovations without expansion. On the other hand, work to expand the built-up area or new constructions must undergo a more careful institutional assessment, which is not carried out by the reserve's administration, but by the central INEA.

All the holiday homes built inside the reserve after its creation are irregular constructions. Most of them have been fined and have had legal actions brought against them, but there are still quite a few that have not yet been fined. This is due to deficiencies in the inspection process, because the inspectors have been there and found no-one. Especially in Pouso da Cajaíba and Ponta Negra, which are the places where we have the most difficulty in fining people."

The Ecological Reserve category is not included in the SNUC. For you, is it considered to be in any group: full protection or sustainable use?

"Strictly speaking, it combines characteristics of both groups, because there is a legal issue that says it is a non-building area, but at the same time it admits the presence of caiçara communities. It's therefore an incongruous situation with the current legislation. This is the situation of the reserve at the moment, it has characteristics of both integral protection and sustainable development. The built-up areas within the reserve are restricted to the water's edge."

I ask: what is the position of the caiçaras in relation to the reserve?

"There are different aspects. More than ninety per cent of caiçaras are aware of the reserve and the fact that it has been created. At the same time, they are not clear about the objectives, what they can and cannot do. They think that the reserve has helped to protect the environment and has helped to protect them from external property speculation. Most of them think that their lives have changed very little as a result of the existence of the reserve. They associate the reserve very strongly with the issue of not being able to build and *not being able to farm. This association is very clear to them.*

I'd like to ask you: what is the organisation's structure like for you and what are its main activities?

"The structure is still not adequate. It falls short of the structure needed to do a good management job. There are only three of us. Since we took over, we have prioritised the following lines of work: re-categorisation, which for us is a watershed and will provide guidance for the management of the unit; the second is the issue of public use and land occupation, at which point we are beginning the study of tourism planning; and the third is the project to implement our "headquarters kit", consisting of a visitor centre, accommodation for researchers, accommodation for park rangers, an administration office, which will allow us to have a more positive interaction with society in general. This is how we've been working."
Testimonial III - Eduardo Godoy: Head of APA Cairuçu.

What is the position of the caiçaras in relation to the Juatinga Ecological Reserve?

"Well, we're going through a process, which has been ten years in the making, to re-categorise the REJ, because it's not a category recognised by the SNUC. So there's a very good team working for INEA, and they've been interviewing and talking to the communities. It's difficult to talk about the caiçaras in the reserve, because each centre there has its own characteristics. So, in general, from what I've seen, from what people have said and from the meetings we've had, everyone recognises and knows that there is a reserve. However, there's a lot of confusion. They don't know what can and can't be done in a reserve, or in anAPA.

The APA Cairuçu management plan establishes a very interesting and innovative figure, which is the Vila Caiçara Expansion Zone. Here, the caiçara

can build, can even expand their villages. As the management plan began in 2005, there are buildings there that date back to before the plan and are not owned by caiçaras, so the person can stay there. It's a very complicated situation, because the caiçaras sell their possessions and the person who buys them usually wants to build. So we have a lot of difficulties enforcing this rule. The caiçaras have ownership there through their ancestors, their parents, their grandparents. However, many people also present land deeds there, and also present themselves as owners of the areas where they are squatters. So many places in Juatinga have land conflicts. It's one of the *main problems there.*

I ask: does the REJ recognise the APA Cairuçu management plan?

The situation is a little more complex. The reserve is administered by INEA. INEA's head office in Rio, the directorates and the coordination, don't really recognise it. This management plan was built considering that it is to be used by the reserve until they have their own. Locally, it's a support. It's a document that guides the team's work. However, it's conflicting, because theAPA brings with it a series of possibilities for construction, for use, for farming. On the other hand, the reserve is treated as an integral protection conservation unit, so none of this can happen. So it's hell.

I ask: what is the relationship like between the two units, ICMBio and INEA?

"It's very integrated. We take part in various activities together, we reinforce their work and they reinforce ours when necessary. Rodrigo and I deal with various issues together.

Here we have a mosaic of units, with an active mosaic council. There are representatives from 16 conservation units, with members from quilombo areas and indigenous areas. It's not just conservation units, there are also traditional populations represented.

Why doesn't INEA consider the APA Cairuçu management plan?

"I don't know. I think it's perhaps a question of the category of the unit, which is treated as if it were full protection and the APA is sustainable use. Maybe that's why, but we can't understand it. Until the reserve has its own management plan, I think it's feasible for them to use ours. Locally, we've managed to work like this, but at the coordination level it's not very well accepted.

5.2.   Visit to Saco do Mamanguá / Juatinga Ecological Reserve

Saco do Mamanguá is an estuarine coastal area in the Atlantic Forest domain, formed by an indentation in the sea approximately 9 kilometres long and 1.5 kilometres wide. It is accessible by boat or by a trail that starts in Paraty - mirim.

According to Diegues (1994), the stories told by the "old-timers" of the area show that it was first occupied by escaped slaves and former slaves, who founded some settlements. In Praia do Cruzeiro, there is still a cross representing the mark of its foundation by a former slave. Other villages, such as Praia do Curupira, have names that refer to popular mythical characters. We can see, therefore, that it is not only material marks that make up these communities, but also symbolic representations.

The purpose of the visit to Saco do Mamanguá was to get to know the caiçara reality *in loco*. To observe their activities, talk to local residents and find out their opinions on the Juatinga Ecological Reserve. We knew that the villages in the area were difficult to reach, so Vila do Cruzeiro was the easiest place to get to from Paraty - mirim.

The journey began at 6am. We had to take a bus from the Paraty bus station to Paraty - mirim and wait on the beach for a boat to show up. Paraty - mirim beach is the place where boats from Saco do Mamanguá and the reserve's northern coast anchor to load or unload goods, and to drop off or pick up people. There are also infrastructures belonging to environmental organisations that demonstrate their ways of using the territory. For example, in front of the final bus stop, there were signs denoting the presence of environmentalist conceptions of the relationship with the ecological environment and trying to inform people that the purpose of the area is environmental preservation. The presence of a park ranger station from the Juatinga Ecological Reserve gave the idea of controlling the territory.

Photo 5 - Paraty - mirim beach with boats starting up (Source: Lucas Grisolia, 2011).

Photo 6 - Plaque in Paraty - mirim (Source: Lucas Grisolia, 2011).

Photo 7 - Park ranger station, Paraty - mirim (Source: Lucas Grisolia, 2011).

The boat used to reach their destination was a motorised speedboat called a voadeira. Two caiçaras, born in Saco do Mamanguá, were unloading empty gas cylinders on the beach to take to Paraty, a secondary activity for both of them when it's closed season and they can't fish. In addition to these activities, they also use their boats to take tourists. When I asked one of them what he thought of the reserve, he replied:

"The *booking doesn't get in the way. I continue to do my work. The only thing is that you have to ask permission from the reserve's management to make your swidden and build (...). I think it's wrong for them to knock down the houses that have already been built there, if they've already allowed the land to be sold why don't they let them build?"* (Caiçara who took me on his boat, Paraty - mirim, 2011).

On the way to Vila do Cruzeiro, you can see a series of holidaymakers' houses and mansions. Some on the tops of hills, others on the edge of beaches. On seeing the ruins of what used to be a house, the caiçara who drives the boat tells us that it belonged to a Korean man who had built it, had his work fined by the agency responsible for the reserve and this resulted in his demolition order. On the other hand, we learn of houses that haven't even been fined.

On arrival, you can see that the beach is a common place for everyone, the site of their daily activities. It's where they socialise. There are fishing ranches, boat moorings and small shipyards where boats are built. One caiçara, who has spent his whole life in Mamanguá, was building a boat by the beach. Very receptive, when asked what he thought of the reserve, he replied:

"*People complain a lot because the caiçara do the farming, but people preserve a lot here. What used to be thatch has become forest. They don't burn it to make crops anymore.*"

Sincerely, he walked a few metres ahead and pointed towards Cairuçu Peak:

"*Look, everything there that's forest used to be thatch, because the caiçara used to set fire to everything in order to make crops. Now, the caiçara know that this is a reserve and they don't set fire to it anymore (...) before, people used to trawl here and there was a lack of fish. Today, it's fine because trawling is banned*" (Caiçara who builds boats, Saco do Mamanguá, 2011).

Photo 8 - Caiçara building a boat, Saco do Mamanguá (Source: Lucas Grisolia, 2011).

The village of Cruzeiro is home to the headquarters of the Association of Residents and Friends of Mamanguá - AMAM. Its entrance staircase begins on the sands of the beach. As well as being the headquarters of the association, the structure also serves as classrooms for caiçara children. Physical education classes are held just opposite. However, they only receive public education up to the fourth grade. The caiçara who drives me tells me that he left Mamanguá and moved to Paraty. One of the reasons was so that his children could attend school. Today, his children live and work in Paraty and he moved back to Mamanguá because things have improved due to tourism.

I know the person who heads the residents' association. According to his accounts, this is where the caiçaras of the village of Cruzeiro and other villages discuss issues that concern the community. He is articulate with other people and mentions his participation in the Forum of Traditional Peoples and Communities. His main concerns are the lack of public education beyond the fourth grade and the lack of public infrastructure (such as electricity and rubbish collection), which doesn't arrive there because it's a reserve. In conversation with his father, a 72-year-old caiçara who had spent his whole life there and used to weave a fishing net right next door, he was told:

"If you *look over there (on the other side of Saco), it's not a reserve. There's electricity there, there's rubbish lorries. The only reason it doesn't come here is because it's a reserve (...) the reserve doesn't help. We want to grow crops to feed our children and we can't. It's only now that they can, after a lot of fighting. Before, people used to come here to trawl. They would come by with their boats and catch everything. It used to be very good here, there were too many prawns, too many fish.*

Now we can't trawl anymore and it's getting better'" (Caiçara weaving a fishing net, Saco do Mamanguá, 2011).

Another caiçara born there is the owner of a small business, Bar do Cruzeiro, which also functions as a small shop supplying basic supplies (such as salt, oil and biscuits) to the community. She says that the trade is good when it's the tourist season and that the shop also supplies basic supplies to some holiday homes when necessary.*"Tourism is very good for us. It brings jobs and business. But there are people who build and don't let us pass along the beach any more. They put fences and dogs on the beach"* (Caiçara owner of Bar do Cruzeiro, Saco do Mamanguá, 2011).

Photo 9 - Caiçara weaving a fishing net, Saco do Mamanguá (Source: Lucas Grisolia, 2011).

Photo 10- Bar do Cruzeiro, Saco do Mamanguá (Source: Lucas Grisolia, 2011).

For another caiçara who also used to weave fishing nets, aged 49, "the *way it is, it's bad. They won't let us build, they embargo the works, they knock down others. We can graze, but before they wouldn't let us"*.

Photo 11 - Caiçara weaving a fishing net, Saco do Mamanguá (Source: Lucas Grisolia, 2011).

The return to Paraty - mirim took place at the end of the day. The caiçara who was driving me still had other things to do with his boat. It then took another hour by bus to reach Paraty.

Understanding how different ways of appropriating the same territory lead to conflictual situations was the main contribution of the visit to Saco do Mamanguá. The Juatinga Ecological Reserve clearly imposes limits on the caiçaras' use of the territory. There is no justification for the fact that public services are not available to residents of the reserve, even though they are available in the area in front of Saco do Mamanguá. The reserve, which has mentioned in its creation decree the objective of fostering caiçara culture, acts by imposing difficulties on their lives. The caiçara child who wants to finish at least primary school is forced to leave his territory. Young caiçara who want to build their own homes and continue to reproduce as such in their own territory come up against contradictory and poorly applied legislation.

The holidaymaker, who is part of an urban-industrial society based on private property and individualism, uses the place to rest from the city environment and enjoy its paradisiacal beauty. He brings with him his form of appropriation/domination of the territory, buying up beaches and fencing them off. The common place of the caiçaras becomes private property, with control and limitations.

For the caiçara, what is his alone is his house, which he built with his own hands. The other places are used by the community in their daily activities, respecting the place used by each individual, such as the gardens, backyards and fishing spots. The caiçara territory has a wealth of meanings that are part of the identity of its people. The stories passed down there and told by the elders, and the symbols, such as the cross nailed to Praia do Cruzeiro by the founder of the village, fill this territory with caiçara life.

Photo 12 - Caiçaras chatting, Saco do Mamanguá (Source: Lucas Grisolia, 2011).

Tourism is intense in the southern region of Rio de Janeiro. It is characteristic of a territory-network that is expanding, giving new uses to the territory. The caiçara, even if they integrate into this new activity, do not lose the characteristics that define them as such, their territory, their symbols and representations. The caiçara who took me on his boat, a resident of Saco do Mamanguá, continues to carry out fishing as his main source of income. He does it as he has learnt in his society, together with his compadres and his typical relationships. He also uses his boats to fulfil demands from his surroundings, such as taking tourists out for a stroll or reselling gas. These secondary activities are also carried out with caiçara characteristics. In this way, working in tourism doesn't mean ceasing to be a caiçara. To leave your territory is to stop being a caiçara.

CHAPTER 6

Final considerations

The geographical conceptions of territory, the ideals of environmental conservation and traditional populations, and the understanding of what protected areas are, were analysed in order to understand the extent to which the creation and implementation of conservation units contributes to maintaining the culture of traditional populations. Based on the bibliography researched, the testimonies of representatives of environmental organisations and the accounts of the residents of Saco do Mamanguá, conclusions can be drawn about how the Juatinga Ecological Reserve interferes in the lives of the caiçaras.

A very important issue is the confusing way in which the law and the decree creating the reserve are applied. By not clearly defining the non-building character of the area, it sets the precedent for various interpretations. Thus, as the head of the REJ himself pointed out, various managers have been able to apply the legislation in the most convenient way for the interests they were contemplating.

The inertia of the environmental agency in complying with what the law stipulates with regard to land regularisation for the benefit of the caiçaras is a strong factor that favours the conflicting situation in the area. As a result, the reserve's land does not legally belong to either the caiçaras or the public authorities, leaving room for speculators. The low level of implementation of the reserve, both the deficiency of the physical structure and the lack of a management plan, contributes to the incongruity of its objectives with the activities that take place within it.

Farming is a characteristic activity of the caiçara. As well as providing food, it also supplies their goods. After harvesting the manioc planted in his fields, the caiçara uses part of it to make flour. This is made in the flour houses, where the caiçara woman is responsible for toasting and then pillaging the food. If the caiçara is forbidden to cultivate, he no longer devotes part of his day to it and the flour house loses its traditional usefulness. In this way, if one of REJ's objectives is to promote caiçara culture, what is being practised is the limitation of this culture.

According to the head of the REJ, Rodrigo Rocha, once the swidden had been abandoned for fallow land, it could no longer be used because it was secondary vegetation in a stage of regeneration, and thus conflicted with environmental legislation. However, Law No. 11.428 of 22 December 2006, known as the Atlantic Forest Law, describes this:

> Art. 23: The cutting, suppression and exploitation of secondary vegetation in the middle stage of regeneration of the Atlantic Forest Biome will only be authorised:
>
> *III - when necessary for small rural producers and traditional populations to carry out agricultural activities or use (...) essential to their subsistence and that of their family (...)*

Therefore, the legislation allows the caiçara to practise swidden cultivation. The ban on this activity in the Juatinga Ecological Reserve is at odds with what is proposed in its decree and has no legal basis.

The construction of the caiçara house is also traditional and involves their marriage practices. Before getting married, the young caiçara builds his house. At this point, the community sees him in a different light and he gains new responsibilities within the group. When he is forbidden to build on his territory, he is forced to leave it. As a result, he cannot reproduce his traditions.

Another latent contradiction is the failure to recognise the APA Cairuçu management plan. Used as a basis locally by REJ managers and not accepted by INEA headquarters in Rio de Janeiro, this shows the contradiction between the interests of the agency and the local reality. The REJ needs a management plan to base its actions on. However, the policy of the management body seems to appreciate the low level of implementation of the reserve and preservationist practices.

The managing body of the Juatinga Ecological Reserve, the INEA, is linked to private interests. Actors such as Paraty City Hall, land grabbers and businessmen have their interests represented in the public body that should be managing the territory for conservation combined with caiçara culture. Thus, what we see is the appropriation/domination of the territory by the state apparatus in the interests of capital, causing the deterritorialisation of traditional caiçara populations.

The Juatinga Ecological Reserve was created in 1992, a time when environmental policies and their institutional apparatus did not take into account the importance of socio-environmental issues. As a result, it has controversial, confusing and poorly applied legislation. INEA relegates the reserve to a low level of implementation, with few employees and little infrastructure to manage it. As a result, it is not managed in the interests of the caiçaras, but for the private sector.

It is well known that today the SNUC takes into account the way of life of various populations through the creation of units that allow the use of natural resources for their traditional activities. It should therefore be pointed out that the issue raised here was analysed in relation to a legally outdated conservation unit with a preservationist stance. Therefore, this same question can be applied to other types of conservation units where traditional populations live.

The caiçaras have their culture limited by the existence of the reserve. They are prevented from farming and building new homes, while at the same time they are under great pressure from speculators and land grabbers. The fact that the area where they live is a reserve serves as an argument for the government not to provide basic infrastructure, which is the right of every citizen. As a result, the exodus is great.

In short, the creation and implementation of the reserve led to the expulsion of the caiçaras from their territory. Therefore, despite its legal objective of fostering caiçara culture, in practice there are direct and important limitations on the way this population appropriates its territory. When the creation of a conservation unit implies interventions in the way a population appropriates its territory, it directly interferes with the maintenance of their culture.

Bibliography

ABIRACHED, C, F, A. Territorial planning and protected natural areas: conflicts between instruments and the rights of populations in Ubatuba - Paraty. Brasilia, 2011. Master's thesis. Centre for Sustainable Development. University of Brasília, Brasília.

BECKER, B. The geopolitics of Amazonia. Estudos Avançados 19 (53), 2005.

Brazil. Resolution 34, of 1 July 2005. Federal Official Gazette. 14 September 2005, Section 1, page 89.

BRAZIL. Constitution (1988). Constitution of the Federative Republic of Brazil. 18. Ed. Rio de Janeiro: DP&A, 2005.

Brazil. Law 6938, of 31 August 1981. Provides for the National Environmental Policy. Federal Official Gazette. Brasília, 2 August 1981.

Brazil. Law 9985, of 18 July 2000. Establishes the National System of Conservation Units. Federal Official Gazette. Brasília, 19 September 2000.

Brazil. Decree 6040, of 7 February 2007. Institutes the National Policy for the Sustainable Development of Traditional Peoples and Communities. Available at: <http//:www.ibama.gov.br>.

Brazil. Law 11428, of 22 December 2006. Provides for the utilisation and protection of the vegetation of the Atlantic Forest Biome. Federal Official Gazette. Brasília, 26 December 2006.

CALLOU, A. B. F. Peoples of the Sea: Socio-cultural heritage and perspectives in Brazil. In Ciência e Cultura, year 62, number 3. 2010.

CARDOSO, E. S. Geography and fishing: contributions for a management model. Revista do Departamento de Geografia, n.14, p.79-88. 2001.

CLASTRES, P. A sociedade contra o Estado: pesquisas de antropologia política. São Paulo: Cosac & Naify, 2003.

DE FRANCESCO, A. A. Territory in dispute: the case of the caiçaras of Cajaíba. In: 5th Anppas National Meeting. Florianópolis/SC, 2010.

DIEGUES, A. C. NOGARA, P. J. O nosso lugar virou parque. São Paulo: NUPAUB/USP, 1994.

DIEGUES, A. C. (org.) Biodiversity and traditional communities in Brazil. São Paulo: NUPAUB/USP, 2000.

DIEGUES, A. C. Povos e mares: uma retrospectiva de socioantropologia marítima. São Paulo: NUPAUB, Série Documentos e Relatório de Pesquisa, n°.9, 1993.

FREITAS, J. G. The Portuguese coast, perceptions and transformations in contemporary times: natural space and humanised territory. Journal of integrated coastal management 7 (2): 105-115, 2007.

GUERRA, A. J. T. COELHO, M. C. N. (Org). Conservation units: geographical approaches and characteristics. Rio de Janeiro: Bertrand Brasil, 2009.

HAESBAERT, R. O mito da desterritorialização: do fim dos territórios à multiterritorialidade. 3ed. Rio de Janeiro: Bertrand Brasil, 2007.

HAESBAERT, R. Territórios alternativos. Niterói: EdUFF; São Paulo: Contexto, 2002.

LOURIVAL, T. D. M. L. Consultancy to establish procedures for characterising and resolving existing conflicts with traditional communities: caiçaras, quilombolas and indigenous people, in the region of Angra dos Reis/RJ, Paraty/RJ and Ubatuba/SP. March, 2009.

MUSSOLINE, G. Aspects of culture and social life on the Brazilian coast. In: Schaden, Egon. Man, culture and society in Brazil. Petrópolis: Vozes. 1972.

PAIVA, M. P. Economic and social situation of artisanal fishermen in Ceará. Bulletin of the Ceará Society of Agronomy. Fortaleza, Vol. 8. 1967.

PARATY CITY HALL. Paraty 2010 participatory masterplan, *Revision of the Paraty municipal masterplan and complementary laws,* Paraty - RJ, November, 2010.

Rio de Janeiro. *State Decree 17.981, of 30 October 1992.* Creates the Juatinga Ecological Reserve. Available at: <http://www.inea.rj.gov.br> .

Rio de Janeiro. *Law 1859, of 1 October 1991.* Authorises the executive branch to create the Juatinga Ecological Reserve. Available at:<http//:www.alerj.rj.gov.br>.

SAQUET, M. A. *Approaches and conceptions of territory.* 2.ed. São Paulo: Expressão Popular, 2010.

SAQUET, M. A. SOUZA, E. B. C. (org). *Readings of the concept of territory and spatial processes.* 1.ed. São Paulo: Expressão Popular, 2009.

STEINBERGER, M. (org). *Territory, environment and spatial public policies.* Brasília: Paralelo 15 and LGE Editora, 2006.

VIANNA, L. P. From invisible to protagonists: traditional populations and conservation units. São Paulo: Annablume; FAPESP, 2008.

ZARUR, G. C. L. (org). Region and nation in Latin America. Brasília: EdUnB and São Paulo: Imprensa Oficial do Estado, 2000.

Printed by Books on Demand GmbH, Norderstedt / Germany